THE MATHEMATICS OF ARTIFICIAL INTELLIGENCE FOR HIGH SCHOOLERS

First published 2023 by Panda Ohana Publishing

pandaohana.com

admin@pandaohana.com

First printing 2023

©2023 Paul Wilmott

ISBN 978-1-9160816-4-2

THE MATHEMATICS OF ARTIFICIAL INTELLIGENCE FOR HIGH SCHOOLERS

Paul Wilmott

Panda Ohana Publishing

Other books by Paul Wilmott

Option Pricing: Mathematical Models and Computation. (With J.N.Dewynne and S.D.Howison.) Oxford Financial Press (1993)

Mathematics of Financial Derivatives: A Student Introduction. (With J.N.Dewynne and S.D.Howison.) Cambridge University Press (1995)

Mathematical Models in Finance. (Ed. with S.D.Howison and F.P.Kelly.) Chapman and Hall (1995)

Derivatives: The Theory and Practice of Financial Engineering. John Wiley and Sons (1998)

Paul Wilmott On Quantitative Finance. John Wiley and Sons (2000)

Paul Wilmott Introduces Quantitative Finance. John Wiley and Sons (2001)

New Directions in Mathematical Finance. (Ed. with H.Rasmussen.) John Wiley and Sons (2001)

Best of Wilmott 1. (Ed.) John Wiley and Sons (2004)

Exotic Option Pricing and Advanced Levy Models. (Ed. with W.Schoutens and A. Kyprianou) John Wiley and Sons (2005)

Best of Wilmott 2. (Ed.) John Wiley and Sons (2005)

Frequently Asked Questions in Quantitative Finance. John Wiley and Sons (2006)

Paul Wilmott On Quantitative Finance, second edition. John Wiley and Sons (2006)

Paul Wilmott Introduces Quantitative Finance, second edition. John Wiley and Sons (2007)

Frequently Asked Questions in Quantitative Finance, second edition. John Wiley and Sons (2009)

The Money Formula: Dodgy Finance, Pseudo Science, and How Mathematicians Took Over the Markets. (With D.Orrell). John Wiley and Sons (2017)

Machine Learning: An applied mathematics introduction. Panda Ohana Publishing (2019)

For my teachers

Contents

PREFACE

If you are reading this, there is still some hope! It is not yet a post-apocalyptic 2029, Skynet has not taken over the world with its indestructible, human-exterminating, Artificial Intelligence cyborgs. Not everything is a depressing shade of grey. If you've seen James Cameron's 1984 movie *The Terminator*, you'll recognize this vision of what might be in store for us if Artificial Intelligence goes rogue. And according to a recent survey, that movie has been so influential that it has informed the popular view of Artificial Intelligence, AI for short, and how the future might evolve.

For the moment though, fortunately, the sun is still shining and the birds are still singing. Even so, there are worrying signs that 2029 is fast approaching, metaphorically speaking. We've heard recently that AI's ability to mimic people's voices, a talent of Arnold Schwarzenegger's cyborg in the movie, is now a reality. (At the time of writing however, time travel is not. But I don't care what Einstein said, if anyone or anything can figure that one out it will be AI.)

Artificial Intelligence is in the news every day. Sometimes in a good way, such as its involvement in the design of new vaccines, but more often for the dangers it poses, manipulation of the news for some evil agenda, or DeepFake videos of celebs in compromising situations. The Pope dressed like a rapper in a Balanciaga puffer jacket was at the silly end of the spectrum but still fooled many people.

But what do we mean by Artificial Intelligence?

AI is the branch of computer science concerning the development of computer programmes and machines with characteristics usually associated with human intelligence. Understanding speech, recognizing faces, diagnosing illnesses, playing video games, driving cars, forecasting stock prices, are all examples of what computers are now able to do that until recently were just the fantasies of science fiction. AI is used in social media and online shopping to recommend the content that you see. It is used to detect banking fraud and weed out spam from your emails. It will tell you that such and such a movie is a '99% fit' to other movies you've enjoyed.

While most, I hope, AI is being designed with good intentions in mind, there are plenty of concerns around the abuses, such as the DeepFake videos. And as artificial intelligence is increasing at such great speed it is probably the case that this is being accompanied by an equally rapid dumbing down of the human species. I am very fond, as a mathematical modeller, of conservation laws and can't help thinking that intelligence across the universe is conserved; as artificial intelligence increases so human intelligence decreases such that the sum of the two remains the same. Am I joking? I've no idea!

And since AI relies on having vast amounts of data there is also the concern about lack of privacy as more and more of your personal data are collected from your viewing, browsing and shopping habits. And while it may seem helpful that a movie is a 99% fit, it does not exactly encourage your exploration of new movie genres. If you like having decisions made for you then you will welcome the new AI world. But if you value your privacy, variety and exploration then maybe you should not rate movies, only use anonymous browsing, and do all of your shopping in brick-and-mortar stores. (And don't forget to pay in cash!)

The concept of AI has been around long before it became a formal scientific field. It can be seen in mythology, literature and old black and white, even silent, movies. Robots are just AI in human(ish) form.

In Greek mythology we have the bronze giant Talos, built by Hephaestus, the Greek god of invention and blacksmithing. A stop-motion Talos created by Ray Harryhausen featured in the 1963 movie *Jason and the Argonauts*, 99% highly recommended (by me, a human of sorts) but sadly currently not available on many streaming services.

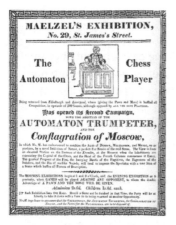

The Mechanical Turk was a chess-playing automaton built in 1770. This impressive machine, created by a man with an equally impressive name, Johann Wolfgang Ritter von Kempelen de Pázmánd, would play against challengers and usually win. It toured Europe and America for over 80 years, being seen by Napoleon Bonaparte and Edgar Allan Poe, but was destroyed in a fire. Later it was revealed by the son of one of the Turk's owners to have been a hoax. A human chess player sat inside the box underneath the chess board and moved the pieces around.

In his 1836 article, Maelzel's Chess-Player, in the *Southern Literary Messenger*, Poe made comparisons between the Turk and the Difference Engine of Charles Babbage. If, *if*, the Turk was genuine then it would be far superior to a mere calculator, he wrote. Poe appreciated the complexity in a game of chess as opposed to doing a few sums. But that was a big if. Poe made it clear that he suspected the Turk to be a hoax.

In the 1927 Fritz Lang silent film *Metropolis*, set in a dystopian future, the automaton, the *Maschinenmensch*, is built by the scientist Rotwang as a dedication to his dead lover. The robot is later transformed into human female form as part of a plan to destroy the city.

And then there's the famous Robby the Robot from the 1956 film *Forbidden Planet*. Robby reprised its role in many TV programmes, earning the epithet 'The hardest working robot in Hollywood.' It appeared in a few episodes of the *Twilight Zone*, another good source of chilling predictions about the possibilities of AI.

And there are many, many more.

In this book my goal is to inform students about the many faces of AI, coming at this exciting subject from different angles. There'll be some history, some mathematics, and some simple projects. But there is a clue in the title of the book, mathematics is at the heart of Artificial Intelligence, and at the heart of this book. You don't need to know any mathematics or computer programming to use AI these days, but if you are really interested in this field, and in its possibilities as well as its dangers, then you will need to at least appreciate the role of mathematics.

Note: When you see text in italics it is often because I am quoting directly from a source, the source being mentioned. I also use the convention of italics for emphasis, for book titles, movies, etc. and for non-English words.

Aside: Projects

On the subject of projects ... whenever I read in a book "Try this at home" or "Photocopy these pages so you can ..." or "Things to make with your children on a rainy day during the school holidays" or especially "Bonding activities to do with colleagues at work" etc. I never, ever, do them. But I also never, ever, feel bad about this. So, if you don't do any of my suggested projects you shouldn't feel bad either. I have written all of them in a way that you can just mentally walk through each stage of each project, for example imagining asking people how tall they are, how much they weigh, then imagining jotting it all down in a book, imagining drawing a graph and so on. Einstein, had his Thought Experiments, *Gedankenexperimenten*, in this book we have Thought Projects, *Gedankenprojekte*.

INTRODUCTION

Definitions

We start with a definition from Wikipedia:

Artificial intelligence (AI) is intelligence—perceiving, synthesizing, and inferring information—demonstrated by machines, as opposed to intelligence displayed by non-human animals or by humans. Example tasks in which this is done include speech recognition, computer vision, translation between (natural) languages, as well as other mappings of inputs.

Another definition from Wikipedia:

Machine learning (ML) is a subfield of Artificial Intelligence that deals with systems that are able to acquire their own 'knowledge' ('learn') by extracting patterns from data instead of having those patterns provided to them directly via programing.

When I describe the mathematics behind AI it is machine learning I will be describing. When we see AI in action, ~~doing your mathematics homework~~ creating a pattern for a sweater you want to knit, we know *what* AI is doing, but we probably have no clue about *how* it figured out how to do what it does. And that's what distinguishes AI and ML from last century's technology. *Learning* is the key. We write algorithms not to teach a computer *how to play* the game of Go but *how to learn how to play* the game of Go.

(By the way, if you do want to use AI to do your mathematics homework note two things: First, that's called cheating, and you will get caught out ... teachers use AI to find out who has been using AI. Head spins! And second, at the moment AI is good at some things, such as making DeepFake videos, but can be really bad at mathematics. See the Aside at the end of this chapter.)

I'm Sorry, But This Is A Mathematics Book!

Yes, it is.

Some of you will be going, 'Woohoo, more mathematics! It's my favourite subject!' But I know for some of you your heart will sink, 'Oh no, more mathematics! It's my *least* favourite subject!'

For the latter, the good news is that there is plenty of non-mathematical material in this book, some history, some nice examples, and discussion of AI matters in general. And at the end of later chapters there are collaborative projects that you can try out with friends. (In the projects, the job for the non-mathematician is not doing the mathematics, no, your job is to collect lots of data in order to train the AI algorithms, such as asking all your friends' parents how old they are and how much they earn. Good luck with that!)

But the bad news is that if you want to make a career in Artificial Intelligence you might need to study mathematics at university. Or maybe computer science. That's if you want to be one of the people creating AI applications. If you just want to use it, maybe to ~~do your mathematics homework~~ create a sweater design, then you'll just need a keyboard and an internet connection.

But, more good news, the mathematics in this book will be quite basic. I will not blind you with science, you will not need to be a whizz kid computer programmer. Everything we do here will just involve pencil and paper.

ChatGPT

I have to mention ChatGPT as it was the first AI tool to become widely used. ChatGPT is an AI chatbot using an algorithm known as a generative pre-trained transformer (GPT). This is a type of neural network (we'll be seeing these in a later chapter) that, after training on large quantities of data, is capable of producing content, including answers to school essay questions, that is hard(ish) to distinguish from human content.

It was launched at the end of November 2022, and had garnered over 100 million users within a couple of months.

While being quite good at producing plausible-looking essay answers it did have a tendency to get its facts wrong. It was also bad at some simple mathematics, such as solving quadratic equations. Even months after launch it was still not able to solve such equations correctly. This is strange since one of the important features of AI is how it can learn. Feedback from all the mathematicians poking fun at ChatGPT's limitations should have resulted in the algorithm improving. But six months after its launch ChatGPT was still making the same mistakes. And six months for AI is probably the equivalent of a hundred thousand years of human evolution!

Aside: Using AI to Solve Your Mathematics Homework. Can You Find the Error?

Me: Solve $x^2 - 8x - 260 = 0$.

ChatGPT: To solve the quadratic equation $x^2 - 8x - 260 = 0$, we can factorize or use the quadratic formula. Let's factorize it:

$x^2 - 8x - 260 = 0$

We need to find two numbers that multiply to -260 and add up to -8 (coefficient of the x term). After considering the factors of 260, we can rewrite the equation as follows:

$(x - 20)(x + 13) = 0$

Setting each factor equal to zero:

$x - 20 = 0$ or $x + 13 = 0$

Solving for x:

For $x - 20 = 0$:

$x = 20$

For $x + 13 = 0$:

$x = -13$

Therefore, the solutions to the equation $x^2 - 8x - 260 = 0$ are $x = 20$ and $x = -13$.

SOME HISTORY OF AI

I've already mentioned a few examples of artificial intelligence from classical mythology, the movies and TV. But these are all rather vague in terms of how the robots, chess player, etc. knew how to do what they do. Were they carefully programmed step by step, or did they learn? It's rather a moot point since none of them was actually real.

Psychohistory

But there's an another literary example, ok still not real, that does hint at how AI might work, and that's thanks to the character Hari Seldon in Isaac Asimov's *Foundation* series of science fiction books from the '40s and '50s. Hari Seldon was a mathematics professor at Streeling University on the planet Trantor. Professor Seldon developed a new field, called psychohistory, that used statistics to predict the future, in a probabilistic sense. Key to this was the use of data from history, vast amounts of data. Whereas it is impossible to forecast the behaviour of an individual, the *average* behaviour might be predictable. An analogy would be the impossibility of forecasting the path of a single raindrop as it is buffeted by wind and other raindrops but if you see it raining outside you have a good idea whether you need to wear a hat. This use of large quantities of data is one of the characteristics of machine learning as we now know it.

The Turing Test And The Birth Of AI

In 1950 the English mathematician Alan Turing, an actual person this time, not fictional, proposed a simple test to judge the realism of any artificial intelligence, as a precursor to answering the question whether a machine can think. The test goes like this, in a text-based conversation between a computer and a human could an observer tell which was which?

Later that decade at a workshop at Dartmouth College in New Hampshire the field of AI, and its name 'artificial intelligence,' was supposedly 'born.' This workshop invited several big names in various related areas to brainstorm on the possibilities of artificial intelligence. Sadly those big names only attended briefly, and not at the same time, so not much was achieved.

Michie And The Matchboxes

Now let me introduce Donald Michie, often called the Father of Artificial Intelligence. Donald Michie had worked on cyphers and code cracking with his colleague Alan Turing at Bletchley Park during the Second World War. After the war he gained a doctorate and in the early 1960s, as a professor in Edinburgh, turned his attention to the problem of training — a word you will be hearing quite a lot of later in this book — a computer to play the game of Noughts and Crosses, a.k.a. Tic Tac Toe. Well, not so much a computer as an array of matchboxes, without the matches. Three hundred and four matchboxes to be precise, laid out to represent the stages in a game of Noughts and Crosses.

You know how to play this game, you have a three-by-three grid on which one player writes an X alternating with the other player writing an O, with each player's goal being the first to get three of their symbols in a row. I imagine you've played this game and already know how to win, or at least to not lose.

You could write some computer code that tells you what is the best play at each stage of the game. This wouldn't be too difficult, just a bit boring. There

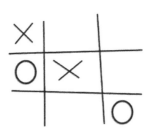

are 304 possible scenarios, positions of the Os and Xs. (Don't forget that there are many symmetries which reduce the number of scenarios.) So you'd have 304 instructions as to what move to make. For example, the code representing the state shown in the sketch on the left would tell us, if we are playing X, to place our X in the top middle (top right would be just as good). Whatever O does, we win on the next move. So that's 304 instructions or lines of code for the computer. Now 304 lines of code is not really all that much. Such a computer code is therefore straightforward for Noughts and Crosses. But once you move onto other, more complicated, games with more states of play and more moves you can make, well, it becomes extremely difficult, maybe even impossible. I'll give an example shortly.

Professor Michie approached Noughts and Crosses in a completely different way. And the beauty of his approach is that it can be used in situations where the traditional programming is too cumbersome.

What Professor Michie did was to take 304 matchboxes, draw on each one a grid representing one of those 304 states of play, and fill the tray of the matchboxes with coloured beads. As we'll see, these 304 matchboxes are really the same as a computer programme.

Here we see what *one* of Professor Michie's matchboxes might have looked like. It's just an ordinary matchbox with a sketch of part way through a game of

Noughts and Crosses. This sketch represents the same state of play as above. Into the tray Professor Michie put lots of coloured beads. In this example the computer is playing the Xs, and it is the computer's turn to play. There are five spaces in which the computer can place its X, how does it decide which? Well, we already know that there are two places to put our X which can lead to a win. These are top middle and top right, in the spaces coloured green and brown. *But we want to avoid telling the computer that's what it should do. We want the computer to figure this out for itself!*

That's what the beads are for. Notice how the colours of the beads are the same as the colours representing the available spaces in the grid. The next move is chosen by closing the tray, shaking the box, and taking out a bead at random. Whatever the colour is, that's the move the computer makes.

So far that's just the computer choosing randomly. The clever bit is coming up …

Put this box to one side, with its chosen coloured bead. Now it is O's turn to play. After their move we are in a different state, and so we go to one of the other matchboxes, the one representing the new state, and again pick a coloured bead at random. There won't be so many colours in this next box as there will be fewer spaces in which to put our next X.

And so on, until the game is over. Then the clever bit …

If we have won the game, we reward all of the matchboxes that were used in this game by putting in some beads of the same colour that was chosen at each state. If we have lost, we take out beads of those colours. If it's a tie we put the one bead chosen for each box back in its tray, that is, nothing has changed.

Eventually the matchboxes fill up with beads representing the most successful action at each state, and the beads representing losing moves disappear. Now when we use the matchboxes to tell us what moves to make it will always pick the best moves at each stage of the game.

Clever stuff!

Do you see how we never *told* the 'computer' what was the best move at each state of the game? Instead, it *learned* the best moves by being rewarded for success and being punished for failure.

The 'machine,' and Michie called it MENACE, for Machine Educable Noughts And Crosses Engine, has thus learned to play Noughts and Crosses. Professor Michie's method was groundbreaking both in its use of what is now known as reinforcement  learning and in its physical representation of a computer programme. Oxford's Associate Professor in Data Science used to make his students re-construct MENACE, as although today's computers are faster and more sophisticated, he thought that MENACE still encapsulated the basic principles of computing.

Deep Blue

Now onto actual computers rather than matchboxes. Deep Blue is a chess programme developed initially by Carnegie Mellon University and later by IBM starting in 1985. In 1995 it played against world champion Garry Kasparov but lost four out of six games. But the next year, after further programming, it beat him, thus becoming the first artificial intelligence machine to beat a world champion.

DeepMind

Next of particular interest is the story of how Google DeepMind programmed a computer to learn to play Go. And in 2016 it promptly went on to beat Lee Sedol, a 9-dan professional Go player. AlphaGo used a variety of techniques, among them neural networks. Again, it learned to play without explicit programming of optimal play. Curiously most of that play was not against human opponents, it mainly played against itself.

To show how difficult Go is, and consequently how difficult was the task for Google DeepMind, if you used matchboxes to represent states then you'd need a lot more than 304 of them, you'd probably have to fill the known universe with matchboxes. I would strongly recommend watching the movie *The AlphaGo Movie* — Is there an Oscar category for Most Boring Movie Title? — to see a better description than the above and also to watch the faces of everyone involved. Lee Sedol started moderately confident, became somewhat gutted, and finally resigned (literally and psychologically). Most of the programmers were whooping for joy as the games progressed. Although there was one pensive-looking one in the background who looked rather troubled. (It might just have been indigestion, or maybe he was worried about where AI could be heading.) At times the computer made some strange moves, moves that no professional Go player would have considered. In the movie there was discussion about whether this was a programming error, lack of sufficient training, or a stroke of computer genius. It turned out to be always the last of these. It has been said that the game of Go has advanced because of what humans have learned from the machine.

Aside: The Game of Go

Go is a board game for two players using a board with a 19-by-19 grid. (Beginners play with a smaller board, 9 by 9 or 13 by 13.) Playing pieces, black and white, called stones, are placed on the intersections of the gridlines. The aim of the game is to surround more territory than your opponent. The game originated in China 2,500 years ago and is possibly the oldest board game continuously played to the present day. I would heartily recommend you learn this game! Even on a 9-by-9 board you will get a sense of the magnitude of the number of positions possible.

The Good News

New York Times, May 25th, 2023, 'A Paralyzed Man Can Walk Naturally Again With Brain and Spine Implants'

Gert-Jan Oskam was living in China in 2011 when he was in a motorcycle accident that left him paralyzed from the hips down. Now, with a combination of devices, scientists have given him control over his lower body again.

It is a colossally difficult job interpreting the many, many electrical signals generated when someone thinks about moving a limb. Which signals are responsible for what effect, which are responsible for movement?

In the new study, the brain-spine interface, as the researchers called it, took advantage of an artificial intelligence thought decoder to read Mr. Oskam's intentions — detectable as electrical signals in his brain — and match them to muscle movements.

The Lancet, April 21st, 2023, 'Development and international validation of custom-engineered and code-free deep-learning models for detection of plus disease in retinopathy of prematurity: a retrospective study'

Lack of experts in the field of paediatric ophthalmology means that it is not possible to diagnose Retinopathy of Prematurity (ROP), a leading cause of childhood blindness, as often as needed. Researchers used an AI programme trained on thousands of images of the eyes of new-born babies. The AI was as successful at finding ROP as the expert ophthalmologists, making diagnosis possible in places where there is a lack of such experts.

Phys.org, September 28th, 2022, 'AI better than humans at detecting blue whale calls'

Using machine learning, a team from the Australian Antarctic Division, the K. Lisa Yang Center for Conservation Bioacoustics at Cornell University, and Curtin University, have trained an algorithm to detect blue whale 'D-calls' [social calls made by whales in feeding grounds] in sound recordings, with greater accuracy and speed than human experts.

The D-calls vary greatly over time, and from whale to whale. This makes analysis difficult, particularly automated analysis. Or so it was thought.

Researchers took 5,000 whale calls from around Antarctica. Six humans classified each of these calls as either a D-call or not. The algorithm was then used on this classified training set, as described later. And then the algorithm was used on a test set of new calls to see how well it did compared with the classification done by human experts. An independent judge, Dr Miller (human!), then determined which classifications, both by humans and by the algorithm, were correct.

And the results were … the AI found 90% of the D-calls, whereas the humans only found 70% of them. The machine had won.

'It took about 10 hours of human effort to annotate the test data set, but it took the AI 30 seconds to analyze this data, 1,200 times faster,' said Dr Miller.

United Nations Environment Programme, November 7th, 2022, 'How artificial intelligence is helping tackle environmental challenges'

More climate data is available than ever before, but how that data is accessed, interpreted and acted on is crucial to managing these crises. One technology that is central to this is Artificial Intelligence.

AI is being used in many, many areas to help solve the climate crisis. To name just a few: Making houses more energy efficient; Monitoring deforestation; Monitoring methane levels; Predicting CO_2 and sea levels.

The map app on your phone helps you to find the best driving route from A to B, and therefore minimize carbon emissions from your car.

Not everything in the garden is rosy though. All this AI analysis brings with it its own environmental costs:

The ICT [Information and Communications Technology] sector generates about 3-4 per cent of emissions and data centres use large volumes of water for cooling.

Which brings us onto …

The Bad News

The Conversation, May 23rd, 2023, 'Is generative AI bad for the environment?'

In 2019, researchers found that creating a generative AI model [one that produces complex outputs such as images, sentences, music] called BERT with 110 million parameters consumed the energy of a round-trip transcontinental flight for one person. The number of parameters refers to the size of the model, with larger models generally being more skilled. Researchers estimated that creating the much larger GPT-3, which has 175 billion parameters, consumed 1,287 megawatt hours of electricity and generated 552 tons of carbon dioxide equivalent, the equivalent of 123 gasoline-powered passenger vehicles driven for one year. And that's just for getting the model ready to launch, before any consumers start using it.

Never mind any of the bad things that AI can be used for, even just the initial building of an AI model for benign purposes is harming the environment thanks to its use of electricity and cooling water.

The Red Hand Files, January 2023, 'I asked ChatGPT to write a song in the style of Nick Cave and this is what it produced. What do you think?'

Verse 1:
In the depths of the night, I hear a call
A voice that echoes, through the hall
It's a siren's song, that pulls me in
Takes me to a place, where I can't begin

Chorus:
I am the sinner, I am the saint
I am the darkness, I am the light
I am the hunter, I am the prey
I am the devil, I am the savior

Verse 2:
I walk the streets, with my head held high
I've got my demons, but I'll make them fly
I'll dance with the devil, and I'll play his game
I'll be the one, to call his name
...

Nick Cave responds:

Dear Mark,

Since its launch in November last year many people, most buzzing with a kind of algorithmic awe, have sent me songs 'in the style of Nick Cave' created by ChatGPT. There have been dozens of them. Suffice to say, I do not feel the same enthusiasm around this technology. I understand that ChatGPT is in its infancy but perhaps that is the emerging horror of AI — that it will forever be in its infancy, as it will always have further to go, and the direction is always forward, always faster. It can never be rolled back, or slowed down, as it moves us toward a utopian future, maybe, or our total destruction. Who can possibly say which? Judging by this song 'in the style of Nick Cave' though, it doesn't look good, Mark. The apocalypse is well on its way. This song sucks ...

Songs arise out of suffering, by which I mean they are predicated upon the complex, internal human struggle of creation and, well, as far as I know, algorithms don't feel. Data doesn't suffer. ChatGPT has no inner being, it has been nowhere, it has endured nothing, it has not had the audacity to reach beyond its limitations, and hence it doesn't have the capacity for a shared transcendent experience, as it has no limitations from which to transcend. ChatGPT's melancholy role is that it is destined to imitate and can never have an authentic human experience ...

Mark, thanks for the song, but with all the love and respect in the world, this song is ********, a grotesque mockery of what it is to be human, and, well, I don't much like it — although, hang on!, rereading it, there is a line in there that speaks to me —

'I've got the fire of hell in my eyes'

— says the song 'in the style of Nick Cave', and that's kind of true. I have got the fire of hell in my eyes – and it's ChatGPT.

Love, Nick

The Guardian, May 24th 2023, 'New Zealand's National party admits using AI-generated people in attack ads'

New Zealand's National party has admitted using artificial intelligence to generate people in their attack advertisements.

The ads included images of a group of robbers storming a jewellery store, two nurses of Pacific island descent, and an apparent crime victim gazing out of a window. One ad even appeared to show the cast of the Fast and Furious *franchise.*

The images, showing a woman with enormous eyes, two nurses with oddly plasticine skin and thieves wearing balaclavas with openings that would not match up to human features, quickly raised suspicions.

Nothing illegal here. And they fairly quickly owned up to using AI. However, it is the thin end of the wedge …

The Guardian, April 6th, 2023, 'ChatGPT is making up fake Guardian articles'

Last month one of our journalists received an interesting email. A researcher had come across mention of a Guardian article, written by the journalist on a specific subject from a few years before. But the piece was proving elusive on our website and in search. Had the headline perhaps been changed since it was launched? …The reporter couldn't remember writing the specific piece, but the headline certainly sounded like something they would have written. It was a subject they were identified with and had a record of covering … they could find no trace of its existence. Why? Because it had never been written.

Luckily the researcher had told us that they had carried out their research using ChatGPT. In response to being asked about articles on this subject, the AI had simply made some up. Its fluency, and the vast training data it is built on, meant that the existence of the invented piece even seemed believable to the person who absolutely hadn't written it.

Forget about bad people using AI for nefarious reasons, here is an example of AI generating its own fake news stories and not telling anyone about it!

And then …

Sky News, June 23rd, 2023 'Lawyers used ChatGPT to help with a case - it backfired massively'

Two New York lawyers have been fined after submitting a legal brief with fake case citations generated by ChatGPT.

Steven Schwartz, of law firm Levidow, Levidow & Oberman, admitted using the chatbot to research the brief in a client's personal injury case against airline Avianca.

He had used it to find legal precedents supporting the case, but lawyers representing the Colombian carrier told the court they could not find some examples cited - understandable, given they were almost entirely fictitious.

Several of them were completely fake, while others misidentified judges or involved airlines that did not exist.

District judge Peter Kevin Castel said Schwartz and colleague Peter LoDuca, who was named on Schwartz's brief, had acted in bad faith and made 'acts of conscious avoidance and false and misleading statements to the court.'

Portions of the brief were 'gibberish' and 'nonsensical,' and included fake quotes, the judge added.

I can easily see AI replacing lawyers. But perhaps not just yet.

Aside: Image Generation, 'Elvis has left the building, ... and is living in a trailer'

I asked Microsoft's Bing to generate a photo of Elvis Presley aged 80 and living in a trailer.

Apart from the *three* misshapen hands (arthritis can explain only so much) and not looking anything like Elvis ... Ok, it's rubbish, but you get what you pay for and this was free. I only included this here for posterity. By the time this book gets onto the bestseller lists I expect photo generation to be indistinguishable from the real thing. When this happens, what will it mean for news reporting, or evidence in criminal cases?

The Three Laws Of Robotics

AI is being criticised, justifiably, for its potential to run amok, especially since it is often impossible to understand what is going on inside the algorithms and since the algorithms are evolving so rapidly. Experts, and worried non experts, are calling for limitations or controls on what AI is allowed to do. Inspiration for such rules could come from another of Isaac Asimov's stories, this time concerning robots. In his 1942 short story *Runaround* he presented the Three Laws of Robotics.

1. A robot may not injure a human being or, through inaction, allow a human being to come to harm
2. A robot must obey the orders given it by human beings except where such orders would conflict with the First Law
3. A robot must protect its own existence as long as such protection does not conflict with the First or Second Law

These could equally be applied to AI.

But then as AI develops, one day it will almost certainly decide it knows what is best for its human 'dependents,' just like every toddler grows up to be a teenager that knows what is best for their parents. So there go the three Rules! (Kids, parents don't really have a clue what they are doing, we just make it up as we go along. I kind of think that's what we are doing with AI as well.)

SOME JARGON, SOME MATHEMATICS AND PRINCIPAL TECHNIQUES

Lagrange multipliers, gradient descent, maximum likelihood, principal components analysis and cross entropy are all subjects you won't be needing. Phew! Not unless you study machine learning at university.

Jargon And Mathematics

For this book you will only need to know a tiny bit about vectors, distance measurement, scaling, cost functions and other simple mathematical concepts. And you'll need to know some jargon. Below are the essentials of this jargon, concepts and mathematics needed for the rest of this book.

Mathematical Model: A mathematical model is a description of a system or process using mathematical ideas and principles. It could be a model of something physical, such as an apple falling from a tree thanks to gravity. Or financial, such as a model representing how your savings are affected by changing interest rates. Biology? How do populations of lions and gazelles increase and decrease? Mathematical models can be used to explain or to predict something. With the example of gravity you can use a simple model to tell how deep a well is. Drop a stone down the well, time how long before you hear a splash in the water. There's a simple mathematical formula that tells you the relationship between time and distance.

Some mathematical models are extremely, *quantitatively*, accurate. Again, gravity. Getting an unmanned spacecraft to Mars is only possible because we know precisely what forces are going to act on the spacecraft due to the various stars, planets, moons, ~~egos,~~ etc.

But some mathematical models are only *qualitatively* accurate. Sometimes we call these 'toy models.' You can't accurately predict the population of lions and gazelles but mathematical models can be used to explain population cycles. The more gazelles the more food for the lions, lion population goes up, gazelles down; Then less food and the lions suffer, while the gazelle population recovers; Repeat.

Some mathematical models are necessarily probabilistic in nature. Radioactive decay, for example.

And there's also the category of models that are accurate but only briefly predictable. These are models of chaos. While the models are quantitatively good, the behaviour depends very sensitively on what are called initial conditions. This is the Butterfly Effect, being the idea, somewhat exaggerated, that a butterfly flapping its wings in the Amazon will affect the weather in Wales, say.

Aside: Are you a numbers person or a symbols person?

Your mathematics teacher writes something on the blackboard, does your heart give a little cheer when they use symbols instead of numbers? Or, maybe you groan inwardly. Do you love all those xs and ys? The more the merrier! Or would you rather stick to three apples, four bananas and a pineapple? Just like there are cat people and dog people, most people are either numbers people or symbols people. I'm very much a symbols person, as long as I know what the symbols mean. The advantage of numbers is that you can't fool people with examples using numbers, if you just use addition, subtraction, etc. Even if someone's mental arithmetic isn't great they will appreciate what you are doing. On the other hand, symbols can easily be used to confuse. However, when you need to see structure in something then you really need to do that in symbols. In real-world machine learning you'll see symbols absolutely everywhere. There'll be a few in this book, but I've tried hard to keep them to a minimum.

Learning: In traditional mathematical modelling you would sit down with a piece of paper and a pencil and ponder, what is it that makes a swing go backwards and forwards in a predictable fashion? There's the force of gravity and the tension in the rope. Let's write down some equations. (With a little help from Sir Isaac Newton.) Or if you wanted to write a computer programme for playing chess then you'd write down some code like 'IF Black Rook something something THEN White Bishop something something.' (I'm not a chess player.) And you'd have many, many lines of IF, WHILE, AND and OR, etc. code telling the programme what to do in complex situations. You might sit on the swing for a while for inspiration, or play chess, not at

the same time though. But mainly you'd just be using your brain. Whatever the problem, you would build the mathematical model yourself.

The above is *not* what happens in machine learning!

In machine learning the approach to modelling is completely different. In machine learning you write a computer programme in such a way that it takes in raw information about the problem at hand, in the form of data, it analyses that data and then the computer code *evolves* so that it can predict future behaviour. In the example of the swing, you might film the swing in slow-mo, then give the computer information about the angle that the swing makes with the vertical and how that changes in time. The final programme would then be able to predict the swing's angle in the future. But at no point would you ever mention the forces on the swing, the laws of Isaac Newton would never be used. Instead, the programme has learned from the data.

Algorithm: An algorithm is a set of rules for someone or something to follow. It could be as simple as the route from your school to your house for use by anyone coming to your party. Or it could be thousands of lines of computer code used in face recognition. When it's the route map to the party we'd usually just call that 'directions' or 'instructions,' it's only really when computers are involved that we use the fancier word 'algorithm.' Running with the map example, you might be trying to get from 10th Avenue and 24th Street in New York City to the Barnes and Noble bookshop in Union Square. The computer algorithm would tell you to walk along the road until you get to a crossroads and then Turn Left/Turn Right/Go Straight Ahead. How the algorithm decides what is the best route to take, which of Turn Left/Turn Right/Go Straight Ahead is best at each junction, would be a nice machine-learning problem. (Even though the streets are laid out in a grid, the traffic lights make this non trivial!)

Dependent Variable or Output: The dependent variable or output is the quantity, factor, thing, you are modelling, trying to understand or forecast. For example, you have a 'ukulele and you are going to pluck a string. Can you tell beforehand what will be the frequency of the sound it produces? The frequency is the dependent variable, or the output, the prediction.

Independent Variables or Inputs: The independent variables or inputs are the quantities, factors or things that affect your output. Back to the 'ukulele, the length of the string, its tension and density would be the inputs. Change the inputs, the length etc., and the output, the frequency, will change. The mathematical model would be the relationship between the inputs and the outputs.

Data: Data is the raw information, often in numerical form. The data will be used by our algorithms for forecasting. Number of books on your bookshelf, length of bananas in the supermarket, age of the parrots in the zoo, and so on. Data can be analysed, manipulated, played around with, each algorithm has its own rules.

Feature: Features are characteristics of the data. They might be numerical or they might be descriptive. Take the example of the features that might be associated with a person. Features could be their age, hat size, number of pets, temperature of their freezer, and so on. So every person among our data has the same type of information, we know all of their ages, hat sizes, etc. In order for the computer algorithm to be able to do calculations we will turn non-numerical features into number form.

Optimization: If written properly, the machine-learning algorithm tries to be the best it can be at forecasting or finding rules. That's called optimization. In the swing example, the algorithm wants to find the smallest difference between what it predicts will happen to the swing in the future and what actually happens. In getting to Barnes and Noble, it wants to find the optimal, the shortest, route.

Supervised Learning: In supervised learning you are given both input and output data for the algorithm to work with. For example, the inputs might be

pictures of dogs that have been digitized with numbers representing the colours in the pixels, as [R,G,B]. The top left of this photograph of a dog becomes the

	A	B	C	D
1	Hound				
2	[255,255,255]	[255,255,255]	[255,255,255]	[255,255,255]	[25!
3	[255,255,255]	[255,255,255]	[255,255,255]	[255,255,255]	[25!
4	[254,255,253]	[254,255,253]	[254,255,253]	[254,255,253]	[25·
5	[253,255,252]	[253,255,250]	[253,255,252]	[254,255,251]	[25·
6	[248,253,247]	[249,254,247]	[250,255,249]	[251,255,249]	[25
7	[246,251,244]	[247,253,243]	[248,253,246]	[250,255,246]	[25!
8	[244,255,241]	[245,253,240]	[246,254,243]	[248,255,243]	[24!
9	[244,252,239]	[245,253,240]	[246,254,241]	[247,255,242]	[24!
10	[243,254,238]	[243,254,238]	[243,254,238]	[243,254,238]	[24·
11	[243,254,238]	[243,254,238]	[243,254,238]	[243,254,238]	[24·
12	[242,255,238]	[242,255,238]	[242,255,238]	[242,255,238]	[24·
13	[242,255,238]	[242,255,238]	[242,255,238]	[242,255,238]	[24·
14	[243,254,240]	[243,254,240]	[244,255,241]	[244,255,241]	[24·
15	[243,254,240]	[243,254,240]	[244,255,241]	[244,255,241]	[24·
16	[243,253,242]	[243,253,242]	[244,254,243]	[244,254,243]	[24·
17	[243,253,242]	[243,253,242]	[244,254,243]	[244,254,243]	[24·
18	[244,255,241]	[243,254,240]	[243,254,240]	[243,254,240]	[24·
19	[243,254,240]	[243,254,240]	[243,254,240]	[243,254,240]	[24·
20	[244,255,241]	[243,254,240]	[243,254,240]	[243,254,240]	[24·

numbers on the right. But then there are also outputs. In this example, the pictures might have each been classified as Hound, Toy, Terrier, etc.

These classes are the outputs. Our algorithm will learn from these pictures and then we use it to classify any new doggy photograph we give it. Is the new dog we see a Hound, etc.? We can also use supervised learning for predicting numerical values for outputs for any new data points we input. For example, input temperature and time of day, and predict household electricity usage.

Unsupervised Learning: Unsupervised learning is when the data are not labelled. This means that we only have inputs, not outputs. In a sense, the algorithm is very much on its own. The algorithm finds relationships or patterns for you. We show the computer digitized pictures of dogs, and the computer groups or clusters them together according to whatever features it deems most important. Perhaps it will also come up with Hound, Toy, Terrier, … or perhaps black, brown, white, … or something completely different that humans don't even notice. Unsupervised learning might at first seem strange, but you'll understand it if I said to you 'Here are the contents of my desk drawer. Please sort them out.' *Supervised* learning would start by labelling contents as 'Things to write with,' 'Things to hold pieces of paper together,' etc. *Unsupervised* learning does not give the machine any clues at all.

Vector: A vector is a list of numbers, such that the position in the list tells us what the number represents. This needs an example to be understood.

What we see here is a column of numbers, inside stretched out parentheses.

$$\begin{pmatrix} 27 \\ 7^3/_8 \\ 6 \\ -21 \end{pmatrix}$$

That's how we write a column vector. (We can also have row vectors, but we won't be using those here.) Don't let the parentheses worry you, they're there just to keep the numbers together! Each data point will have such a feature vector. In this example, the

$$\begin{pmatrix} 30 \\ 7^1/_4 \\ 5 \\ -18 \end{pmatrix}$$

feature vector for Person A (above left) might have the person's age as the top number, the second hat size, and so on. Person B (to the right) would have a similar feature vector but with their own numbers in each entry.

Distances: We'll often be working with such vectors, and we will want to measure the distances between them. The shorter the distance between two feature vectors the closer in character are the two samples they represent. There are many ways to measure distance. Here are two of the most common.

Euclidean distance: The classic measurement, using Pythagoras, just square the differences between vector entries, sum these and square root. This would be the default measure, it's distance as the crow flies. Using the numbers in the two vectors we see above, we'd get the following:

$$\sqrt{(27 - 30)^2 + (7.375 - 7.25)^2 + (6 - 5)^2 + (-21 - (-18))^2} = 4.36069.$$

Manhattan distance: The Manhattan distance is the sum of the differences between entries in the vectors. The name derives from the distance one must travel along the roads in a city laid out in a grid pattern, like much of the island of Manhattan. This distance measure is simpler to calculate than the Euclidean distance, there's no squaring and squarerooting. For our vectors here:

$$(30 - 27) + (7.375 - 7.25) + (6 - 5) + (21 - 18) = 3 + 0.125 + 1 + 3 = 7.125.$$

Scaling the Data: The observant among you might have noticed that the distances measured above, with both the Euclidean and the Manhattan measures, are dominated by the first and last, the top and bottom, entries in the vectors. The difference in hat sizes, if that's what the second entry is, makes only a relatively tiny contribution to the distance between the two vectors. But that's clearly silly if you are a hatmaker using AI to help design hats! We need to tweak the data, the numbers in the vectors, so that the hat sizes become just as important as the other numbers.

Let's do a simpler example, with just two features, so vectors have only two entries. Suppose we have data for three people, their heights and weights. And suppose that height is measured in feet and weight in pounds. Person A might be five feet, weighing 180 pounds, Person B six feet and weighing 170 pounds, and the third, C, is also six feet tall but 190 pounds.

$$\begin{array}{ccc} A & B & C \\ \begin{pmatrix} 5 \\ 180 \end{pmatrix}, & \begin{pmatrix} 6 \\ 170 \end{pmatrix}, & \begin{pmatrix} 6 \\ 190 \end{pmatrix} \end{array}$$

(These are obviously rather extreme numbers, but good for demonstration purposes if not for their health.) Using the Manhattan distance measurement, because it's the simplest, the distances between these vectors are

$$AB = (6 - 5) + (180 - 170) = 11, BC = 20, CA = 11.$$

It looks like B and C are nothing like each other, the distance between them, 20, is the largest out of the three distances. Even though B and C are both six feet tall, and for their heights, very slender. This doesn't seem to make sense. The problem is that because the heights are small *numbers* they don't have much impact on the distances. But a person's height is very important when comparing people's physical characteristics. We need to rescale.

Rescaling makes height and weight *both* important distinguishers of physical appearance. Rescaling can be done in several ways. One is to add a number to all heights and divide by another number so as to make the range of heights go from -1 to +1. You do the same with all the weights, but the number you add and the number you divide by would be different.

Ok, you are probably wondering what it means to have negative numbers for heights and weights. Well, there comes a point in all mathematics where you have to stop worrying about *physical* meaning and just work with concepts and numbers. And this is that time! Distance between vectors, which is what we are working with here, doesn't care whether the numbers in the vectors are positive or negative. (If you are still worried think about the two common temperature scales of Celsius and Fahrenheit. To get from one to the other is similar to the rescaling that we have here, but no one worries that a positive Fahrenheit could be a negative Celsius.) However, if it really, really bugs you to have the negative numbers then you could choose to rescale to have a range of zero to one instead. It doesn't matter, but I do rather like the symmetry of -1 to 1. Now back to our rescaling of heights and weights.

Let's first rescale the heights. To all three heights subtract 5.5 and multiply the result by two. So the 5 becomes $2 \times (5 - 5.5) = -1$, and 6 becomes $2 \times (6 - 5.5) = 1$. That's the heights done. Now similarly for the weights. But now subtract 180 and then multiply by 0.1. 180 becomes 0, 170 becomes -1 and 190 becomes 1. The rescaled vectors are shown here. (I've renamed them as well, A', B' and C'.) Rescaling is rather like using different units for measuring, but units

$$A' \quad B' \quad C'$$
$$\begin{pmatrix} -1 \\ 0 \end{pmatrix}, \begin{pmatrix} 1 \\ -1 \end{pmatrix}, \begin{pmatrix} 1 \\ 1 \end{pmatrix}$$

that can be negative as well as positive. The Manhattan distances for these rescaled vectors are

$$A'B' = 3,\ B'C' = 2,\ C'A' = 3.$$

Now B' and C' are closest, the new distance is just 2 compared to the distances of 3 for A' to B' and A' to C'. And this makes a lot more sense.

Dimension: I use the word 'dimension' quite a bit. You are probably used to talking about the three dimensions of space and one of time, thanks to Einstein and *Back to the Future*. I don't usually mean those. I mean the number of entries in a vector or the number of different variables in a model or what we need for drawing or plotting. For example, the first vector I introduced above has four dimensions, there are four entries, one on top of the other, in the column. When I mentioned the 'ukulele string, the independent variables determining its frequency were the length of the string, its tension and density, that's three variables and a three-dimensional model. But if I wanted to *plot* how this frequency depends on the variables, I'd need to be able to draw in *four* dimensions, one for the frequency, the dependent variable, and one for each of the independent variables. Traditionally we'd plot the dependent variable, here the frequency, on the vertical axis and the independent variable along the horizontal ... uh, oh, we can't plot *three* horizontal axes, can we? And this is why most of my examples have few dimensions, just so I can draw some pictures!

Classes: Classes are different categories into which the data can be divided. I mentioned this above with types of dog, Hound, Toy, Terrier, etc. Often you will

be dealing with objects with characteristics that aren't well captured by numbers. This can present problems when trying to make everything numerical for the computer.

Socks. How could you assign numbers to socks of different patterns? Plain = 1? Spotted = 2? Striped = 3? Does that work? No! Suppose we did that and then presented a new sock to the algorithm and asked it to tell us whether it is plain, spotted or striped. The numerical output might be 2.4. Does that mean the sock is mostly spotted but a bit stripey? That doesn't make any sense since there is no natural numerical ranking of patterned socks.

The computer needs numbers to do its calculations, but when those numbers are an output representing a class or category they don't necessarily mean anything other than just being a label. I'll show you how we get around this problem with the next example.

Here's a subtle one. Suppose you want an algorithm that looks at images of handwritten numbers and tells you what the number represents. You'd first digitize the image, meaning maybe look at each pixel in the image and assign a number to it depending on how grey that pixel is. The algorithm then does its magic and gives an output. You'd think that outputting a single number, say 3.2,

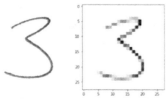

would be ideal, it would be the algorithm's best forecast. But 3.2 does not mean probably a three but a bit fourish. No! We are trying to *classify* images. That the images represent numbers is of secondary importance. Instead, you would output a vector with ten entries. The first entry would represent the probability of the new image being a one, the second entry would be the probability of the handwritten image being a two, and so on up to the last entry being the probability of being a zero. And that's also how we deal with classifying socks, a vector output with as many entries as there are types of sock. The entries in the vector output tell us what the algorithm thinks are the probabilities of any new sock being in each of the categories of plain, spotted, etc.

Parameter: A parameter is a quantity, usually a constant, that appears in a mathematical model. For example, if we think that a straight line can represent the relationship between the amount of dog food bought by a family and the number of dogs they own then a parameter would be the slope of the line. The *form* of the model, the straight line, is always the same regardless of the parameter but the outputs clearly aren't. A simple machine-learning model such as a regression might only have two or three parameters. But something complicated such as a neural network might have hundreds, thousands, or even billions! When I earlier mentioned that the algorithm *evolves*, I really meant that the parameters in the model change during its training in order to achieve an optimal solution perhaps by minimizing a ...

Cost Function: A cost function or loss function is used to represent how far away the predictions of a mathematical model are from the real data. Think of it as a measure of model error. One adjusts the mathematical model, by varying the parameters within the model, so as to minimize the cost function. This is where optimization, mentioned above, comes in. Once the algorithm has been optimized this is then interpreted as giving the best model, of its type, that fits the data. There'll be a fairly simple cost function in the chapter on regression methods later.

Underfitting and Overfitting: Underfitting is when your algorithm is just no good at forecasting. Maybe your algorithm is not suitable, you've been using a decision tree when you should have been using reinforcement learning for example. Or maybe it hasn't learned enough from the data, the optimization is too slow or you don't have enough sample data. Overfitting is the opposite.

Your algorithm seems to do a great job with the data from which it has been learning but it is useless when it comes to new situations. Think of a two-year old who has been putting together the same jigsaw for several weeks. Have they learned how to do jigsaws or how to do the *one* jigsaw? So, give them a different jigsaw. Can they put it together? If they can then they have successfully learned how to do jigsaws. If they can't then they have overfitted, they can only do the jigsaw they have learned from. And that's not useful learning at all.

Training and Testing: Most machine-learning algorithms need to be trained. That is, you give them data and they look for patterns, or best fits, etc. They know they are doing well when perhaps a cost function has been minimized. But you have to be careful that the training is not overfitting. You don't want an algorithm that will give perfect results using the data you've got. No, you want an algorithm that will do well when you give it data it hasn't seen before. When doing the training stage of any algorithm you could use all of the data. But that doesn't tell you how robust the model is. To do this properly try dividing up the original data into training and test sets. Do this randomly, maybe use 75% for training and then the remaining 25% for testing. How well does the trained algorithm do on the test data, the data it hasn't seen before?

Principal Techniques Covered In This Book

There are many different machine-learning techniques. They all come with different types of mathematics, can be used in different situations, and some are more useful than others. For the rest of this book we are going to look at some of the most interesting, and particularly those for which it is possible to do some implementation yourself without being a computer programmer.

Nearest Neighbours: Nearest neighbours, or *K* nearest neighbours to give it its full name, is a supervised-learning technique in which we measure distances between any new data point and the nearest *K* of our already classified data and thus conclude to which class our new data point belongs. In the chapter on this method I start off with a very silly example that describes this perfectly.

Regression: A regression is usually a relatively simple mathematical relationship between inputs and outputs. A straight line relating rhubarb yield and amount of manure, for example. It might be a good model, I don't know. But what about the effect of rainfall? Try a linear relationship in two independent variables, the inputs, the amount of manure and rainfall. But with still the one dependent variable, the output, being the amount of rhubarb produced. Sunshine? Ok, add a third independent variable, and so on. Sometimes you'd need something more complicated than a straight line. If you wanted to relate a person's reported happiness, on a scale of one to ten, to their income then a straight line would be no good ... a straight line wouldn't be confined to the range one to ten. You'd need a different mathematical function for such a regression. Going back to the classes mentioned above, you might use one of these other functions if you wanted to relate the probability of a fruit being an apple to the weight, circumference, colour, etc. of the fruit. Probabilities are always positive numbers lying between zero and one, so again a straight line would be no good.

Clustering: Clustering methods group together similar items. It's up to the computer to decide how best to group things together, we don't give it any information about the data, no clues as to how to do the grouping. So they are unsupervised-learning methods.

Decision Trees: This is a supervised-learning technique. A decision tree is just a flowchart. 'How many legs does it have? Two, four, more than four?' Four. Next, 'Does it have a tail? Yes or no.' Yes. And so on. Like a game of 20 Questions. But can the machine learn and improve on how humans might classify? Is there a best order in which to ask the questions so that the classifying is done quickly and accurately?

Neural Networks: A neural network (sometimes with 'artificial' in front) is a type of machine learning that is meant to mimic what happens in the brain. If an electrical signal along a nerve is too weak it is ignored, if strong enough then a signal gets passed on. In the *artificial* neural network, inputs in the form of a vector are multiplied by a parameter and another parameter is added. This is repeated with different parameters, possibly many times. This gives us a new vector of numbers. Then we take these new numbers and modify them again. This time, as a simple example, take every negative number and throw it away replacing it with zero, and every positive number becomes a one. This is what happens in the brain, the signal is ignored if too weak (negative) or passed on if strong enough (positive).

Reinforcement Learning: In reinforcement learning an algorithm learns to do something by being rewarded for successful behaviour and/or being punished for unsuccessful behaviour. The above-mentioned MENACE was a simple example of this, the rewards and punishments being the addition or subtraction of beads to or from the matchboxes. Computers learning to play board games or video games from the 1980s often involve reinforcement learning.

Now for some techniques in detail! Get out your pencils and graph paper!

IN DEPTH: NEAREST NEIGHBOURS

Nearest neighbours is a supervised-learning technique, nicely explained by a silly example. You walk into a room during a party. You don't know anybody there. People are standing around chatting. You go over to some people and introduce yourself. You notice that out of the five other people nearby, your 'neighbours,' four are wearing glasses. You deduce that you are probably wearing glasses, or, being more sophisticated, that there is an 80% chance, four out of five, of you wearing glasses.

Is this a silly example? Do you think that you might be drawn to specific people because of what they are wearing? Quite possibly. Maybe it's not such a silly example.

This method is commonly called K nearest neighbours (KNN) with the K denoting the number of nearby people or data points that we examine, 1, 3, 5, 21, 51, etc. It would usually be an odd number so we never get a tie. However you don't have to have just two classes (spectacles, none), you can have more. In this case having an odd number of neighbours becomes less important.

What It Is Used For

- Credit risk: It can be used to predict credit risk by comparing an individual's credit history to other credit histories in a database
- Medical diagnosis: It can be used to diagnose medical conditions by comparing an individual's symptoms to other symptoms in a database
- Recommendation systems: The algorithm can be used to recommend products or services based on the similarity of one user to other users
- Pattern recognition: It can be used to classify images, text, or speech based on their similarity to labeled examples
- Data mining: It can be used to find anomalies or outliers in a dataset based on their distance to other data points

- Financial markets: It can be used to predict the future price of financial assets, such as stocks and currencies, based on historical trends and the trends of similar assets
- Intrusion detection: The algorithm can be used to detect malicious or unauthorized activities on a network or system based on their deviation from normal behavior patterns

The Details

K nearest neighbours is perhaps the easiest machine-learning technique to grasp conceptually. Although really there is no 'learning' at all!

We have data, in the form of vectors (coordinates for position in the room, say) for each person, and each data point is classified (such as glasses, no glasses).

Let's do another example, now with some actual mathematics. Is an email spam? The first entry in the feature vector for an email might be the number of words in the email, the second entry might be the number of exclamation marks, the third might be the number of spelling mistakes, and so on. Let's walk through such a spam-identification algorithm.

$$A \quad B \quad C \quad D$$
$$\begin{pmatrix} 193 \\ 4 \\ 0 \end{pmatrix}, \begin{pmatrix} 229 \\ 3 \\ 16 \end{pmatrix}, \begin{pmatrix} 80 \\ 2 \\ 1 \end{pmatrix}, \begin{pmatrix} 616 \\ 0 \\ 0 \end{pmatrix}$$
Spam!

I'll make up the data, and for simplicity we'll just have four emails, *already classified* as spam or not. I'll have three not spam, and one spam. Here we see data from the four emails, *A*, *B*, *C* and *D*. Email *A* has 193 words, four exclamation marks but no spelling mistakes. And so on. We know that *A*, *C* and *D* are not spam, only *B* is a spam email. (The observant will notice a lot of spelling mistakes, the bottom row in each vector, seems to suggest an email is spam. But the point of artificial intelligence is to do this algorithmically, and to be honest I have made this rather obvious.) This is our 'training data.'

Along comes a new email represented by the next vector. Is it spam or not?

New
$$\begin{pmatrix} 98 \\ 4 \\ 2 \\ ? \end{pmatrix}$$

Well, it's got a few spelling mistakes, but on the other hand as a fraction of the length of the email it's not that many. Hmm.

To do the comparison properly we need to rescale the training data, as explained earlier, so that the numbers across the first entry, across the second, and across the third entries are all in the same ballpark. That way all three entries will be important in the analysis. If we didn't do this then the number of words in an email would be the only determinant as to whether or not an email is spam.

I've chosen to rescale so that the smallest number in the first row becomes -1 and the largest becomes +1. The mathematics behind this is simple. Notice that the largest number across the first row is 616 and the smallest is 80. So the transformation for the top row goes like this:

$$A: 193 \rightarrow \frac{2}{616-80} .193 - \frac{616+80}{616-80} = -0.58$$

$$B: 229 \rightarrow \frac{2}{616-80} .229 - \frac{616+80}{616-80} = -0.44$$

$$C: 80 \rightarrow \frac{2}{616-80} .80 - \frac{616+80}{616-80} = -1$$

$$D: 616 \rightarrow \frac{2}{616-80} .616 - \frac{616+80}{616-80} = +1$$

This is a more complicated version of the rescaling I did earlier with heights and weights. But the complication is only because the numbers are messier.

Now you need to rescale the second row, using the minimum and maximum numbers of zero and four. You should be able to figure out the transformations for row two using these new numbers. Then the third row in each vector is rescaled, using the minimum and maximum values of zero and 16.

Finally, also rescale the new, unidentified, vector using the same three transformations.

Here are those four vectors suitably rescaled, together with our new, mystery, email.

$$
A' \quad B' \quad C' \quad D' \quad New'
$$
$$
\begin{pmatrix} -0.58 \\ 1 \\ -1 \end{pmatrix}, \begin{pmatrix} -0.44 \\ 0.5 \\ 1 \end{pmatrix}, \begin{pmatrix} -1 \\ 0 \\ -0.88 \end{pmatrix}, \begin{pmatrix} 1 \\ -1 \\ -1 \end{pmatrix} \quad \begin{pmatrix} -0.93 \\ 1 \\ -0.75 \end{pmatrix}
$$
$$
\quad\quad\quad\quad Spam! \quad\quad\quad\quad\quad\quad ?
$$

Now it's fairly clear that our rescaled new vector is closest to the first rescaled vector, now labelled A', and we conclude that it's *not* spam. I confirmed this by measuring the distances between the new data point and the training points using Pythagoras. See the table. I'm not expecting you to measure distances using Pythagoras, although in a spreadsheet all of the above is rather simple.

In practice there would be a few differences from this simple example here:

Email	Distance to New'
A'	**0.43**
B'	1.88
C'	1.01
D'	3.28

- We would have a lot more training data. Many thousands of emails or more
- We might have more features, so our vectors would have more than three entries
- When we have a new data point to classify we wouldn't look at just the nearest data point from our training set, we'd look at maybe the nearest three, five, 11, 101, etc. We would decide on the best K by putting aside some of our training data for testing purposes, to see if we can correctly identify them using our remaining training data. Inevitably we will misclassify some, but we choose the K to minimize that inaccuracy

Mathematically, the trickiest part of the analysis is the rescaling. In the Project below I have a cunning way of doing this without any calculations.

A Real Example

I have lots of data for the weights and heights of adult males and females. Here are some of the data in a spreadsheet. (See the Data at the end of the book for the original source.) Heights are in inches and weights in pounds.

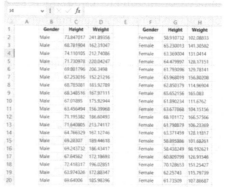

The data are then shown in the plot. I've scaled the heights and weights, as discussed. The green

dots are the males and the yellow are female. The horizontal axis are the rescaled heights and the vertical the rescaled weights.

I've also made sure that there are the same number of males as females, so as to not skew any results for new, unclassified, points.

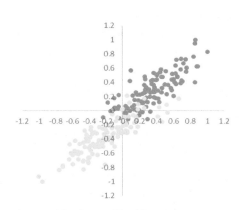

This is a nice example for KNN, there are only two features so we can easily draw pictures. Once we are in three or more dimensions plotting becomes difficult, and in many real-world problems there could be 100s of features in the vectors.

This plot can be used to identify the gender of a new person by putting their data point on this plot and looking at its nearest neighbours. (Don't forget to rescale the new data point in the same way as the training data has been scaled. For the record, the data I used had a range of heights 56.2 to 75.2 inches, and weights 84.9 to 241.9.)

Look at the next picture, it's a close-up of the previous plot. There's now also a red dot in the picture. This is a new, *unclassified*, datapoint. I've drawn two

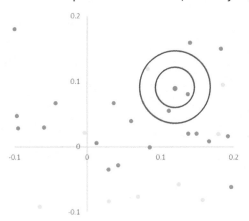

circles centred on the dot. The inner circle encompasses the first nearest neighbour. This neighbour is male. We might therefore conclude that the red dot is a male. But I've also drawn a larger circle, encompassing *three* neighbours. (Remember that it's best to have an odd number to avoid ties.) Of these three neighbours, two

are male and one female. This is interesting statistically. Having more points makes our statistics more robust, but not all being the same gender means that we are less sure that the new data point is male.

The Project for this chapter is producing your own version of this plot.

Your Project

You are going to be repeating this height and weight analysis. You will collect height and weight data from many adults, each data point will be classified as either male or female. You will then take new unclassified data points and see if you can predict which ones are male and which are female. This project requires some obvious sensitivity.

1. Ask family and friends, parents and teachers, to tell you their height and weight. To save any potential embarrassment you should give them a randomly generated ID number! Collect as many as you can.

Remember I said you must rescale otherwise distances could be meaningless? Well, I have a trick that will save you a lot of calculations.

2. Get some graph paper. Mark out the largest square that you can. The paper will probably be rectangular, but I want you to work within a *square*.

3. From your data, note which is the lowest weight from all of the data. Lump all the data together for this. I don't want the lowest weight among males and lowest among females, I want lowest among *all*. Now look through the data to find the heaviest. Do the same for height, the shortest and tallest among all of the data. For this it doesn't matter what units you use, feet and inches or centimetres, pounds or kilograms.

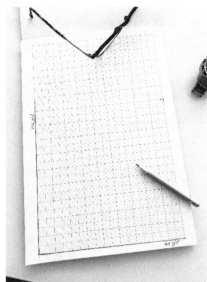

4. On the horizontal axis mark the left end as the shortest height, and the extreme right of your square as the tallest.

5. On the vertical axis mark the lightest weight at the bottom and the heaviest at the top of the axis.

6. Now add a dot on the graph for each person in your data set. Use one colour for males and another colour for females. Using the square and having the horizontal and vertical ranges matching the extremes of the data automatically scales the data in the required way. No mathematics required!

7. It is time to predict whether someone new, someone not in your sample data, is male or female. Get their height and weight.

8. Mark this data on your graph. Don't use one of the two colours, you haven't predicted the gender yet, use a pencil and mark the paper gently.

9. Take a pair of compasses. Centre on the new dot and mark out a circle that captures the nearest five, say, neighbouring dots.

10. Count how many males and females are inside this circle.

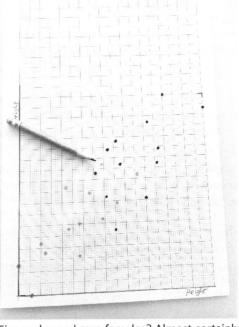

11. You can use your results for forecasting. Five males and zero females? Almost certainly your new sample is male. Three to two? Probably male, but not so sure.

12. How do the results change if you draw larger circles, capturing more neighbours than five?

If you feel uncomfortable asking people their heights, weights and gender then choose two other features, and different classes for analysis. I suggest just two features because then it is easy to draw the necessary plots. But it doesn't have to be just the two classifications.

IN DEPTH: REGRESSION

Regression methods are supervised-learning techniques. They try to explain a dependent variable in terms of independent variables. The independent variables are numerical and we fit straight lines, polynomials or other functions to predict the dependent variables. The method can also be used for classification, when the dependent variables are usually zeroes and ones.

You might know about regression from fitting straight lines through a set of data points. For example, you have values and square footage for many individual houses, is there a simple linear relationship between these two quantities? Any relationship is unlikely to be *perfectly* linear so what is the *best* straight line to put through the points? Then you can move on to higher dimensions. Is there a good linear relationship between value and both square footage and number of garage bays?

In terms of mathematical complexity regression methods are a little bit harder than what we've seen so far. The first time one sees a regression one might draw a plot of the dependent variable on the vertical axis against a single independent variable on the horizontal axis and draw a straight line through them just using judgement and common sense. However, to do regression properly, especially when there is more than one independent variable and plotting is impossible, we will need to minimize a cost function as mentioned earlier.

What It Is Used For

- Advertising: Regression models can be used to estimate the impact of different variables on the outcome of a campaign, such as retail sales or marketing campaigns, and optimize the resources accordingly
- Customer trends: The models can be used to understand the behavior and preferences of customers or users based on their attributes and interactions, such as on streaming services or e-commerce websites

- Data mining: The models can be used to discover patterns and relationships among variables in a dataset, such as finding correlations, outliers, or anomalies
- Pattern recognition: Regression models can be used to classify images, text, or speech based on their features and labels, such as face recognition, sentiment analysis, or speech recognition
- Financial Industry: They can be used to understand the trends in stock prices, forecast prices, and evaluate risks
- Environmental Science: The models can be used to predict the impact of climate change based on various factors
- Psychology: Regression analysis can be used to predict the likelihood of depression based on various factors such as age, gender, and socioeconomic status. It can also be used to predict the likelihood of substance abuse based on various factors such as family history and peer pressure

The Details

In the simplest case we plot dependent variables against independent variables and then draw a straight line through them. That just means that the horizontal, x, coordinate is the feature, and the vertical, y, coordinate is the quantity that we are trying to predict. The dots might be numbers of pupils in a classroom on the vertical axis, and the grade or year on the horizontal.

The question is then just what is the *best* straight line through the dots?

It can get much more complicated than this. The relationship might not be captured well by a straight line, for example growth of bacteria in a Petri dish. At least it's still possible to plot the quantity of bacteria against time in a two-dimensional graph.

But sometimes there is more than the one independent variable. Take house prices against square footage, number of garage bays, and distance from the nearest train station. That would require a four-dimensional plot. The vertical, dependent, axis would be the house prices and you'd need three horizontal, one for each of square footage, number of garage bays and distance from the train station. Well, we can't plot in four dimensions!

For this book, we aren't going to go anywhere near problems for which we can't draw the pictures so we'll stick to two dimensions. And when we get to your project you'll only need to fit a straight line.

Back to the question, what is the best straight line, what is the mathematics behind this?

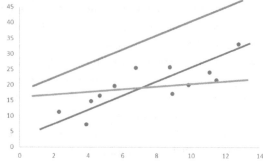

Look at the picture. Which of the three straight lines looks like the best fit to the dots?

It's meant to be obvious that the blue line is the best. The red is too high, all the data points are below it. The green at least has dots above and below but it has the wrong slope.

This illustrates the idea that we need to determine the height of the line and its slope, the two parameters in any equation for a straight line. And these will be chosen to make the difference between the dots and the straight line as small as possible. As mentioned, this difference is usually described by a mathematical expression called a cost or loss function.

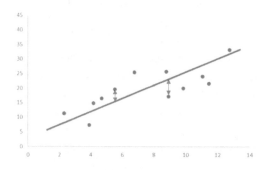

Looking at the next plot you can see I've drawn a couple of distances, one of the dots being above the blue line and one below. The most common cost function takes every one of these distances, one for each dot (12 in this example), squares these numbers and adds them all up. That gives a measurement of the error. We then vary the height and slope of the line so that this error is made as small as possible.

I'll go through the mathematics now for an example with just three of the dots from the data in the above plots. The data will be (2.3, 11.3), (3.9, 7.4) and (4.7, 16.6). (I could have used all the dots from my data but that would just have meant the mathematics would be a lot longer. Not different. Just longer!)

And let's call the line $y = ax + b$.

Taking the xs and ys from our three data points, the error I've described is then the cost function

$$(2.3a + b - 11.3)^2 + (3.9a + b - 7.4)^2 + (4.7a + b - 16.6)^2.$$

As we change a and b the value of this changes. This might be time for you to learn about spreadsheets, if you don't know already! (See the Aside later.)

Playing around with these parameters we find that the best fit, the choice of parameters giving the smallest total error, is

$$y = 1.54x + 6.15.$$

Doing this with more data points, more dots, is the same in principle. Just the result would be different, of course.

A Real Example

In this example I am going to use a simple regression to find the acceleration due to gravity. And I'm going to make life harder for me than for you with your project. I'm going to fit a quadratic function, a parabola.

A ball thrown directly upwards will rise with *decreasing* speed until it stops and falls back down with *increasing* speed. This is all thanks to gravity. There is a simple experiment, involving taking height measurements, from which we can calculate the value of the acceleration due to gravity. In a nod to Sir Isaac Newton, I could have used an apple instead of a ball.

These are the steps in this experiment.

1. I needed a way of measuring height. It can't be too low or the numbers wouldn't be accurate. So I measured out metres along a piece of string, marking them with paint in a bright colour. I hung this string from a second-floor window so it reached to the ground.
2. I positioned a phone with slow motion capability far enough away so that the bottom and the top of the string were within the screen.
3. I set the phone to slow motion, pressed record, and my assistant threw the ball so that it got close to the highest point marked on the string. We did this a few times.
4. I chose one of the recordings to study.

5. I went through the slow-motion recording and wrote down a list of times and heights. I've measured heights from the point at which my assistant let go of the ball. At this point the 'times' are as measured on the phone, I had to do a bit of googling to find out how these times convert to real time, as in seconds. Frames per second and all that.

6. Now I got out my graph paper and drew two axes. The horizontal was time, or movie frames, and the vertical was the height of the ball as measured by the string. Ok, I'm fibbing, I did this in a spreadsheet not on graph paper.

7. Transfer the points to the graph paper/spreadsheet. If you do this you should have something that looks like the picture below. It can be a bit tricky getting accurate distances and timings. But don't worry too much about accuracy. This book is about the mathematics of machine learning, not about video editing.

Time	Height
0	0
0.01	1.4
0.02	2.6
0.03	3.2
0.04	4.2
0.05	4.7
0.06	4.8
0.07	4.4
0.08	4.3
0.09	3.2
0.10	2.0
0.11	0

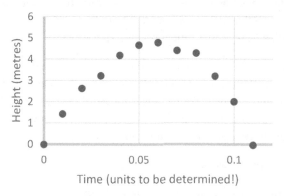

Time (units to be determined!)

8. Now comes the technical part, finding the best parabola to match these dots.

If I had asked you to fit the best straight line then that would be easy. You'd get out a ruler and just draw a line which looked best, maybe with about half the dots above the line, and half below. But the curve I want to fit is a quadratic. That means a formula of the form

$$\text{Height} = at^2 + bt + c.$$

With t being 'time.' The goal is to find the three parameters, a, b and c that optimize the fit. But first, an observation.

This example is rather unusual in the world of machine learning. There is something special here that won't happen very often. Let me explain.

We have some dots on graph paper representing our data. We want to fit some curve to these dots, that's our mathematical model. What shape do those dots look like? Well, I guess they look rather parabolic. So let's fit a parabola. At this point in most problems we might get a reasonable fit. If we didn't we'd have to try a curve of a different shape, cubic, quartic, something even more complicated perhaps. That's what would often happen. But our data are from a very specific problem a ball rising and falling under gravity. And according to Sir Isaac Newton, the curve should be *exactly* a parabola! In other words, we have chosen a function for the regression that happens to be exactly what science tells us it should be!

Believe me, such a happy coincidence only happens in textbooks, and exams.

According to Newton, the relationship between the height and time is the parabola but with

$$a = -\frac{1}{2}g .$$

Here g is the 'acceleration due to gravity.' It has the value 9.81 m/s^2.

Now back to our experimental data and finding the best-fit parabola. Since we are measuring time from zero and height of zero we can set the parameter c to zero, so there's only a and b to be found.

This is where we need the cost function, that measures how far our model (a parabola here) is away from the data (the dots). As mentioned, several times, the cost function that is often used in regression problems is the sum of all the squares of the errors at each point, this is the quadratic cost function,

$$\text{Cost Function} = \sum_{i=1}^{N}\left(\text{ActualHeight}_i - \text{ModelHeight}_i\right)^2 .$$

In this N is the number of data points, and in this example N = 11, I'm not including the point (0,0). The **ActualHeights** come from the data, and the

ModelHeights come from the formula for the parabola. For example, for the first dot we have

$$\text{ActualHeight}_i = 1.4$$

$$\text{ModelHeight}_i = a\, 0.01^2 + b\, 0.01.$$

When you add up all the terms in the Cost Function you get some expression with unknown a and b. Then you must find the values of these parameters that minimize this function. And that's called Ordinary Least Squares (OLS).

Let's not worry how that minimum is found. It's quite easy in a spreadsheet (see the Aside at the end of this chapter), there's even a formula that gives the minimum, or you could just play around with values until the OLS error seems small. In the Project below I'm not expecting you to do anything too complicated, here it's enough for you to know that mathematically it's quite a straightforward problem.

I bet you're wondering how close my result was to the correct value. No? Anyway, from my data and the best-fit parabola I found the acceleration due to gravity to be 9.6 m/s². Quite good, I thought. (Don't forget that the 'time' in my data was not real time. I had to do a conversion from phone slow-mo time to real time. If you repeat this experiment you will have to do the same, and that means finding out how many frames per second your particular app uses.)

Here's my original data again, now with the best fit parabola.

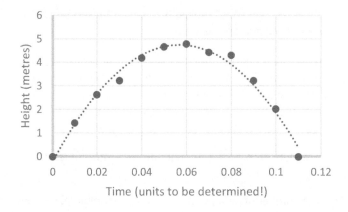

Your Project

An easy one. Is there a relationship between the weight of a dog and the amount it eats? Actually, why restrict this project to dogs? Let's include any and all pets that you have.

This is what you and your friends need to do.

1. First weigh your animals. Possibly easier said than done. If it's a cat or a small dog then that should be quite straightforward, just weigh yourself and then weigh yourself while holding your pet. The difference is your pet's weight. Dog too heavy? Your vet might have a record, perhaps it's in their pet passport. If your dog is not a mix of breeds then google will have the answer. Rabbit, guinea pig? Kitchen scales! Perhaps put it in a bowl first! Mouse, rat, getting a bit small even for kitchen scales, and they do tend to be a bit fidgety. Google it. Budgie? Straight to google, don't even bother trying.
2. Weigh the food given to your pet in one day. Not too many problems here, at least not for the straightforward pets. But more than one dog eating out of the same bowl? Perhaps treat as one large beast? Might require some thinking about. Snake? I think we might skip snakes as they don't eat frequently. Spiders, lizards, stick insects? Might skip them as well. Tropical fish? Definitely for another time.
3. Plot daily food consumption against weight, one dot on your chart per pet. I suggest that you do a different graph for each type of animal.
4. Now I want you to fit a straight line to these dots, one straight line for each animal. And that's your regression.

Notes:

You don't need many animals per chart. You don't need hundreds, not even dozens. But I definitely want you to have more than two or three. The problem with having just two points for a straight-line regression is that the fit will be perfect. This might give you a false sense of confidence in your fit, the error will be exactly zero. The more dots you have on the graph the more you will get a feel for whether or not a straight line is a good model.

Since you are not going to be measuring distances between feature vectors, you don't need to do any rescalings.

Don't worry about doing the fit mathematically. I've explained the idea of a cost function, and minimizing it. But I'm really not expecting you to go to such lengths. I'll be happy with an eyeballed fit.

5. Having done your plot, one for each animal, I want you to measure two numerical quantities from your line, first the intercept with the *y* axis and, second, the slope of the line.

6. The *y* intercept is where your straight line crosses the *y* axis. It's how much you would feed your pet if it weighed nothing! Obviously that's a bit of a silly concept, but it does tell you whether a straight line is going to be valid for lower weights.

7. The slope is measured by drawing a triangle as shown and calculating the ratio of the height of this triangle to its width. To get the most accurate value you should draw as large a triangle as you can. How does this slope vary from animal to animal?

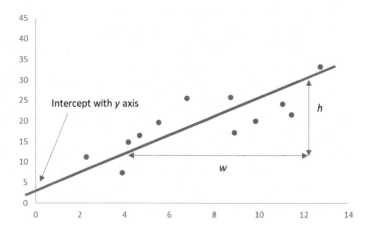

Aside: Using a spreadsheet for regression

If you want to minimize a cost function using Excel there are a few simple steps.

1) Input your data (under x and y below) 2) Put guesses for the two parameters, a and b, into two cells (the yellow cells below) 3) Calculate the distances from data to line 4) Sum the squares (the blue cell) 5) Optimize, using Excel's inbuilt Solver.

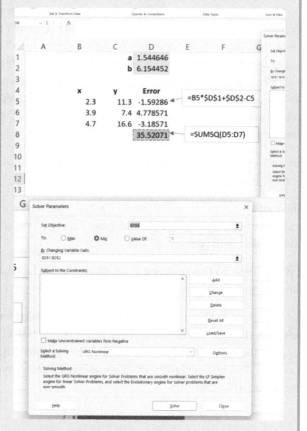

The formulae are shown in the figure. You'll also have to enable Excel's Solver. In Excel go to Home -> Options -> Add-ins.

But really don't go to these lengths unless everything else in this book has been too easy.

IN DEPTH: CLUSTERING

I visited a stately home recently, a National Trust property near the south coast of England. It had the usual distinguished family portraits on the walls, the ancient oak beams, noisy plumbing. They had rooms dedicated to cheese and beer making, as you do. And, of course, there was the library. I've seen many such rooms in similar properties, but this time I thought I'd ask the tour guide a question, 'How did the previous owners organize their bookshelves?' Now I was expecting an answer along the lines of 'Fiction on this wall, non fiction on that.' Or, 'Classics here, modern there.' But, no, the answer was, 'There was a fire a couple of decades ago, all the books were destroyed, so the owner went to an antique shop in London and bought old books by the yard. This is just how they came out of the boxes.'

Fair enough, that's one way to organize your shelves. (But it did make me wonder, where the family portraits had come from!)

And that's what clustering is about. You can imagine the many ways you might group your books: Fiction, non; Old, new; Colour of the covers; Size; Personal rating; Value; and so on.

If you were to do this grouping yourself then you'd be labelling each book like this, and this would then be a classification problem with *supervised* learning.

But clustering is when the algorithm does the grouping for you. The data would be the type of information I've just listed, fiction or non, age, colour, etc., but then the algorithm is let loose to group items together in the way that it sees fit. This makes clustering an *unsupervised* technique. And that makes clustering particularly fun.

What It Is Used For

- Recommendation systems: The algorithms can be used to group users or items based on their similarity and preferences, and then recommend products or services that are likely to interest them
- Search Engines: Receive a mix of results that match your original query
- Classifying documents: The algorithms can be used to organize documents into different categories based on their topics, keywords, or sentiments
- Medical diagnosis: These algorithms can be used to find patterns and anomalies in medical data, such as blood pressure, heart rate, or symptoms, and help diagnose diseases or conditions
- Market segmentation: Clustering algorithms can be used to divide customers into different segments based on their demographics, behavior, or needs, and then design marketing strategies for each segment
- Sports Science: Use clustering to identify players that are similar to each other
- Image segmentation: Clustering algorithms can be used to partition an image into different regions based on their colour, texture, or intensity, and then perform tasks such as object detection, face recognition, or compression

The Details

Just like the previous Nearest Neighbours algorithm, clustering methods use distances between feature vectors to measure their similarity to each other, the smaller the distance between vectors the more similar are the samples.

But clustering methods, of which there are several, differ in usually being unsupervised techniques, so data points are not labelled or classified.

There are several ways in which similar vectors are grouped. Here are three.

K-Means Clustering: *K*-means clustering is an unsupervised-learning technique. We have lots of data in vector form, think dots in space. And we have *K* points or centres of mass in this space. Each data point is associated with its nearest centre of mass, and that defines to which category each data point belongs.

Then we find the optimal position for these centroids that gives us the best division into the *K* classes.

Take a look at the picture. The small dots are the data, the feature vectors are just two dimensional, hence just the two axes. It's clear that there are two distinct groups, those towards the top right and those towards the bottom left. The blue and red dots are the centres of mass of each group, the data points closest to blue dot are in one group and those closest to the red are in the other. These centres of mass are found by iteration. That there are two groups is obvious in this example. In practice, things won't be that obvious, especially when the feature vectors have more than two dimensions!

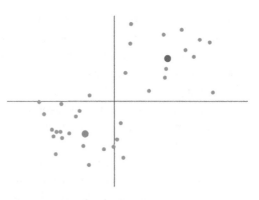

Density-based Clustering: In this method data points are grouped according to the density of points. The clusters are separated by zones in which density is low. There is a lot of flexibility in the shape of the boundaries in this method.

Self-organizing Maps: A self-organizing map is an unsupervised-learning technique that turns high-dimensional data into nice, typically two-dimensional, pictures for visualizing relationships between data points. Imagine a chess board and into the squares you put data points with similar characteristics. You don't specify those characteristics, they are found as part of the algorithm, 'self' organizing. I'll give an example of this method next.

A Real Example

Let's look at an example of self-organizing maps. This is taken from the voting records of the UK's House of Commons. But before looking at the results I want to explain why self-organizing maps (SOM) are particularly useful.

In all of my examples and in all of the projects, I have focused on cases for which we can draw nice pictures. And that means I have restricted myself to at most two-dimensional feature vectors. Such simplicity is rare in real-life machine learning. Feature vectors can be any length, running into hundreds or thousands of entries per data point. I've no idea how much data Facebook holds on everyone but I'm sure that the resulting feature vectors will be humungous.

Let's suppose we are analysing a problem with 100 features. We obviously can't plot in 100 dimensions. Sure, we can output results of analyses, we can output distances between vectors, we can say who is in one group, who is in another, etc. but we can't *visualize* the results easily.

That's where SOM is clever.

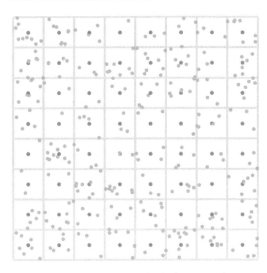

This image illustrates what happens with SOM. Remember that the feature vector can have any number of entries, it need not be two dimensions. So don't be fooled by the two-dimensional nature of this plot!

The green dots correspond to the data points. The red dots are 'representative vectors.' The green dots are put in the box with the representative vector to which they are closest. You can see that some boxes have lots of green dots. And some have very few. Now we can easily see which samples are similar because they are close to each other in the grid, and those which are far apart are very different.

But where do these representative vectors come from? They are found by an iterative technique that's a bit beyond the scope of this book. That's the 'self' in self-organizing map.

Now our real example.

In the Data section at the end of this book I tell you where you can find information about the results of voting in the House of Commons. The data shows which way each Member of Parliament (MP) voted in each vote. Think of each MP as having a feature vector with, say, a hundred or so entries. The top entry will be something like the first vote on such and such a date, the second will be the next vote, and so on. If they voted Aye then the entry is +1, if Noe then -1, if they didn't vote then zero. For each MP the entries represent exactly the same voting occasions.

There are 650 MPs. And we want to put them and their 100-dimensional feature vectors in a two-dimensional plot, to see how similarly or otherwise they voted.

Well, the results, as shown here, are fairly predictable! The heights of the surface in this plot are the number of MPs in each box in the grid. There are really just two types of MP, corresponding to the two main political parties. You can just about see from the plot that there are a few brave MPs who don't vote 100% with their party colleagues, but not many of them!

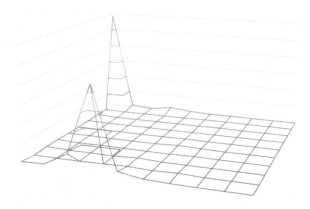

In the Project you won't be doing anything as complicated as SOM, but you'll get close.

Your Project

In this Project you are going to be collecting the sort of data that Netflix collects. Netflix uses this data to decide what movies or series to show and also which movies and shows to recommend to individual users. This Project requires a classroom of students, say 20, who will each label a set of movies as ones they enjoyed or didn't. Someone, a teacher perhaps, will have a specific, but very simple, role to play.

1. The teacher makes a selection of 10 or so movies that they and the students will probably have seen.

2. These movies should encompass many different genres: Thriller; Superhero; Animated; Romcom; Musical; Period drama; etc.

Movie	-1/0/+1
Matilda	
Frozen	
E.T.	
Mary Poppins	
...	

3. Print out one marking sheet, as shown, for every student.

4. Each student then privately rates each movie as +1, if they liked it, -1 if they didn't. And zero if they have no opinion or haven't seen the movie.

5. At the bottom of the sheet each student should also write down three movies that are not listed but which they particularly enjoyed

6. The teacher does the same on their sheet.

7. The teacher must now complete three more sheets with representative answers. Perhaps one sheet has +1 for all the animated movies and -1 for the rest, perhaps +1 for all the musicals and -1 for the non musicals, and perhaps +1 for all the action movies, -1 for the rest. These three sheets together with the teacher's own are going to be the representative vectors to which we shall be measuring distances.

8. At this point no one will have seen the students' answers.

9. The teacher holds up, or writes on the blackboard, the four representative vectors (not yet revealing which one is theirs). These are the A, B, C and D clusters. Each corner of the room is then labelled as A, B, C or D.

10. Each student now measures the distance from their own vector to each of the four representative vectors. Just use the Manhattan distance for this as it is simplest. With 10 movies, the total distance

for each person from the representative vectors could be as low as zero, if they exactly match a representative vector, or as high as 20, if they had exactly the opposite opinion for every movie.

11. Each student now knows which representative vector is closest to theirs. It is the one with the shortest distance.
12. Students then move to the corners of the room with their closest representative vector, to the closest cluster.

And discuss! Compare vectors/answers with people in your cluster. Are close friends in the same corner of the room? What do the sporty students think of Musicals? Are some corners much more popular than others?

And reveal which of the four corners is the teacher's vector.

Now look at the three movies at the bottom of each students' sheets. Are there similarities between students in the same corner?

Finally, homework for the weekend, although I don't know whether the teacher will approve, should be for the students to watch one of the movies recommended by someone in their corner, one they haven't seen before.

IN DEPTH: DECISION TREES

Decision trees are a supervised-learning technique. They are very like flowcharts, or a way of representing the game of 20 Questions.

In 20 Questions the trick to winning is to figure out what are the best Yes/No questions to ask at each stage so as to get to the final answer as quickly as possible, or at least in 20 or fewer questions. In machine learning the goal is similar. You will have a training set of data that has been classified, and this is used to construct the best possible tree structure of questions and answers so that when you get a new item to classify it can be done quickly and accurately.

One small difference from 20 Questions, the questions in a decision tree do not have to be binary as in Yes or No, and the answers can be numerical.

First some jargon and conventions. The tree is drawn upside down, the 'root' (the first question) at the top. Each question is a 'condition' that 'splits' features or attributes. From each node there will be 'branches,' representing the possible answers. When you get to the end of the path through the tree so that there are no more questions/decisions then you are at a 'leaf.'

Building or growing a tree is all about choosing the order in which features of your data are looked at.

What It Is Used For

- Data mining: Decision trees can be used to discover patterns and rules in large and complex datasets, such as customer segmentation, market basket analysis, or fraud detection
- Decision making: Decision trees can be used to visualize and analyze different scenarios and outcomes based on a series of choices, such as medical diagnosis, investment planning, or game strategy
- Feature selection: Decision trees can be used to identify the most important features that contribute to the prediction of the target variable, and eliminate irrelevant or redundant features

- Explainability: Decision trees can be used to provide interpretable and transparent explanations for the predictions made by the model, as they can be easily visualized and understood by humans
- Pattern recognition: Decision trees can be used to classify images, text, or speech based on their features and labels, such as face recognition, sentiment analysis, or speech recognition
- Sports: Decision trees can be used to predict the outcome of games
- Education: Decision trees can be used to predict student performance

The Details

I've found a list of the all-time most popular actors. It's on the website of YouGov, the survey company, the URL is in the Data section at the end of this book. I'm going to pick one of the top 100 and you have 20 questions to find out who I'm thinking of. Should you first ask whether they are alive or dead, or are you better off asking whether male or female? Or just go straight to picking individual names?

I'll give you some information to help you decide what first to ask, that is, what question is to be at the root of your decision tree.

Out of the 100 actors, 70 are male and 30 female. (I didn't compile this list!) Out of the 100, 78 are alive, 22 dead. (If I'd made the list I would have chosen more from Hollywood's Golden Age!)

Which is the best question to ask first?

If you go for the alive/dead question you could be lucky if the response is dead, in which case you have immediately eliminated a large 78% of the possibilities. But on the other hand, you'll probably (that same 78%) only eliminate a mere 22%. Hmm.

With the male/female question, the numbers are 70% and 30%. So, 70% of the time you eliminate 30% of the possibilities.

With this sort of problem it can be beneficial to use extreme examples to gain insight. (That's a very common technique in mathematics.) Suppose you go straight to asking, 'Is it Audrey Hepburn?' You'll almost certainly be wrong, and will have wasted one question, with almost no progress. That is, 99% of the time you eliminate a mere 1%. Continue guessing individual names and you'll probably use up all of your 20 questions and still not have guessed correctly.

Inspired by all this, if it's between asking male/female or alive/dead, or picking names, you should ask, 'Is the actor male?' There's more information in the answer to this question. 'Information' has a technical mathematical meaning, but I'm not going into that here. There's even a whole subject called Information Theory, originally developed by juggler, unicyclist, and, oh yes, mathematician, Claude Shannon.

There's actually an even better question, but this is far from obvious. If you ask, 'Does their name include the letter L?' Then there are exactly 50 that include L and 50 that don't. That's very useful information, 100% of the time we rule out 50% of possibilities. If we can rule out half of the actors with each question then we'll have the final answer in no time! Well, in seven or eight guesses actually. The mathematics of this is to follow.

By the way, I know it was an odd question to ask, is there an L in the name. But for teaching purposes I had to find some characteristic of the actors that would give me an equal split. I was lucky that I found this one. (With a little help from Excel, of course.)

There's a lot of mathematics behind decision trees. One day, if you study this subject in detail, you'll need to know about logarithms, entropy and information theory. But not today.

Oh, I *was* thinking of Audrey Hepburn, by the way!

A Real Example

Here's some more of the mathematics that was hinted at in the 20 Questions example above.

I'm thinking of a whole number from one to 32. How might you go about figuring out what the number is? You can ask me any questions you like.

Is it three? No.

Is it 27? No.

14? No.

This could take a long time.

You could be lucky guessing like this, you might get it first time. Or you might be equally unlucky and get it the 32nd guess. How long would it take on average?

With questions like this it is often helpful to start with smaller number.

Suppose I'm thinking of either one or two. There's a 50% chance, ½, of getting it right first time. And an equal chance of taking two guesses. This gives an average of ½ x 1 + ½ x 2 = 1 ½.

Now take the case of numbers one, two and three. This will take an average of two guesses if you pick one number each guess. Which makes sense. Four numbers, it's 2 ½. With n numbers it will take an average of ½ $(n+1)$ guesses. With 32 numbers it will take an average of 16 ½ guesses.

Can we do any better than this?

Suppose we have numbers one to four. If we ask whether the number is two or less, and we get the answer 'No' then we've narrowed it down to numbers three and four. At that point we guess it's three. If we are right then we've got it, if wrong we now say four and we are done. So that is now an average of 2 ½. That's exactly the same as if we'd picked numbers at random. But instead of taking from one to four guesses, now we need a minimum of two and a maximum of three.

Repeat this idea with numbers one to 32, and this splitting technique gives us an average of 5 ½, with a best case of five guesses and a worst case of six. As opposed to the picking-one-at-a-time method with an average of a relatively enormous 16 ½ with a best case of one and a worst of 32.

Part of the decision tree for this algorithm is shown here.

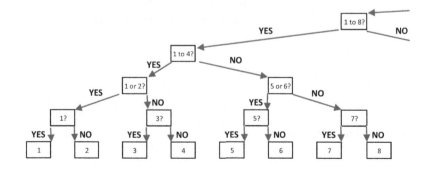

The next graph shows how the average, best and worst cases depend on the range of numbers. It is clear that random picking is not the way to go, unless you are feeling extremely lucky!

Here's a question for you, what is the range from which we can isolate a single integer, using this bisection technique and 20 questions?

Easy, 2^{20} = 1,048,576. Wow!

Your Project

And now for a short relaxing break after all your previous hard work.

I want you to play a few games!

But obviously the games are related to the decision-tree machine-learning technique.

Hangman

The first game to play is Hangman. Choose letters to complete a word. If the letters are in the word, they are written in the correct place. If not, then a sketch of a hanging person is built up line by line. To win this game you need to have an idea of the frequency of letters in the alphabet.

In his 1843 short story, *The Gold-Bug*, Edgar Allan Poe wrote, '*Now, in English, the letter which most frequently occurs is e. Afterwards, the succession runs thus: a o i d h n r s t u y c f g l m w b k p q x z. E however predominates so remarkably that an individual sentence of any length is rarely seen, in which it is not the prevailing character.*' The main character in that story used this list of frequency to translate a coded letter.

This ordering is not quite accurate, however. According to more recent analysis the order, in English, is e t a o l n s h r d l c u m w f g y p b v k j x q z. Such information is useful when playing hangman as it gives at least a few good choices for initial guesses. You don't start with z or q, rather e, t and a. Thereafter, you would use clues about commonality of letter pairs, placement of vowels, etc. You can easily imagine a decision tree for the game of hangman, although it would not be pretty! One subtlety is that it helps to know how cunning your opponent is. Do they also know, and are exploiting, the frequency of letters?!

What might the decision tree for Hangman look like? What would the top, the first guess, be? And how large would the whole tree be?

Wordle

In a similar vein, the *New York Times* Wordle is a game in which you have to work out a hidden five-letter word. You have six attempts to guess the word, feedback is given for each guess in the form of coloured tiles indicating when letters match or occupy the correct position. Each guess has to be an actual word, it can't be random letters. There are 12,972 words that can be used for the guessing but there are only 2,315 possible final answers.

Again, a decision-tree approach is possible.

Wordle

Is there a best word for the first guess? One that gives the most useful information? According to *Scientific American*, popular first choices are AUDIO and ADIEU because they contain many vowels. Online you can find plenty of discussion about the optimal choice for the opening guess, and how it all relates to information theory and entropy. One suggestion is the word SALET, meaning a light round helmet extending over the back of the neck. But there is also discussion of how knowing this spoils the game (and it's not really practical anyway unless you have the full decision tree to hand so you know what the best *second* guess is!).

Mastermind

The final game I want you to play is Mastermind, a two-person codebreaking game from the 1970s. The Mastermind board consists of ten or so rows of four holes. Into the four holes at one end of the board one player, the codemaker, puts four coloured pegs. These are hidden by a plastic shield from the other player, the codebreaker. The original game has six different colours, and depending on how complicated you want to make the game you can insist that the holes have pegs of different colours, or colours can be repeated. Just as in Wordle there will be clues for the codebreaker to try to figure out what pegs are behind the screen.

The codebreaker makes a guess by placing coloured pegs in the row nearest them. The codemaker then indicates how many pegs are the correct colour and in the correct hole, or correct colour but the wrong hole. This is done with further pegs and holes to the side of the rows containing the guesses. Black for

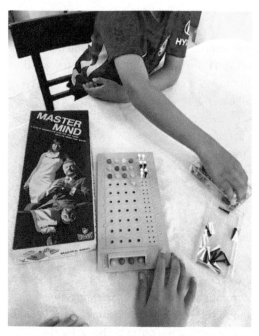

the correct colour in the correct hole, white for the correct colour but in the wrong hole.

Naturally, there is a lot of mathematics behind this simple game.

With four pegs, six colours and repeats allowed, there are 6 x 6 x 6 x 6 = 1,296 possible patterns.

Play this game now a few times, to get a sense of what a realistic number of tries it might take to break the code.

According to Donald Knuth, it can always be solved in a maximum of five goes. But you have to follow his strategy. The first guess needs to have two colours duplicated. So if, say, blue and yellow are two of the six colours, you should start with pegs Blue, Blue, Yellow, Yellow. Of course, it doesn't matter which two colours, it can be any two but they must be repeated in your first guess. After that it gets complicated!

His strategy, which is just a decision tree, is based on the minimax principle. This principle is a decision rule used for minimizing the possible worst-case scenario. It's not trying to be best on average. Instead, it is being pessimistic about future guesses.

IN DEPTH: NEURAL NETWORKS

An artificial neural network is a type of machine learning that is meant to mimic what happens in the brain. Mathematically speaking, numerical signals are received by 'neurons' where they are mathematically manipulated and then passed on to more neurons, repeat. The input signal might pass through several layers of neurons before being output, as a forecast regression or classification. You'll soon see some pictures that explain all this.

Neural networks can be used for either supervised or unsupervised learning.

What It Is Used For

- Image recognition: Neural networks can be used to identify objects in digital images. This has a wide range of potential applications, such as security, search engines, medical diagnosis, face recognition, or self-driving cars
- Speech recognition: Neural networks can be used to convert speech signals into text or commands, which can enable natural language processing, voice assistants, speech translation, or speech synthesis
- Natural Language Processing: They can be used to analyze and generate natural language texts, which can enable tasks such as sentiment analysis, text summarization, machine translation, chatbots, or text generation
- Data mining: They can be used to discover patterns and insights from large and complex datasets, which can enable tasks such as anomaly detection, fraud detection, customer segmentation, or recommendation systems
- Forecasting: Neural networks can be used to predict future values of a variable based on its historical trends and other factors, which can enable tasks such as stock market prediction, weather prediction, demand forecasting, or sales forecasting
- Pattern recognition: They can be used to classify data into different categories based on their features and labels, which can enable tasks

such as spam detection, credit risk assessment, or handwriting
recognition

- Feature extraction: Neural networks can be used to extract
meaningful features from raw data, which can enable tasks such as
dimensionality reduction, feature selection, or feature engineering

The Details

Neurons in the brain communicate with each other using both electrical and
chemical signals. When a neuron receives inputs from one or more other
neurons it will pass on this signal
to other neurons only if the
signal it has received is strong
enough. If it's not strong enough,
then the received signal is
ignored.

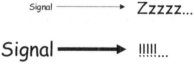

Signal ⟶ Zzzzz...

Signal ⟶ !!!!!...

That's what happens in one neuron. But the human brain has over 80 billion of
them! And each neuron might be connected to 10,000 other neurons!

This response of a neuron to a signal can be represented by a schematic
diagram showing the
mathematics behind this.

The incoming signal, our
input, is x, and the outgoing
signal, the output, is y. A
positive x counts as a strong signal, negative as weak.

X ⟶ Is x > 0 ?

Yes ⟶ y = 1

No ⟶ y = 0

In an *artificial* neural network, the ones used in machine learning, there is a bit
more mathematics going on. First of all, the input x is modified a bit, by
multiplying by some constant parameter, a, and then adding another
parameter, b, so it becomes x'.

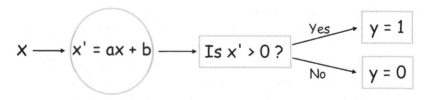

$x \longrightarrow x' = ax + b \longrightarrow$ Is x' > 0 ?

Yes ⟶ y = 1

No ⟶ y = 0

But then there might be a vector of inputs ... $x_1, x_2, x_3, ...$

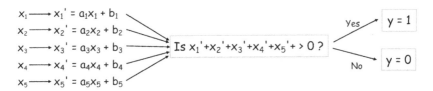

$$x_1 \longrightarrow x_1' = a_1x_1 + b_1$$
$$x_2 \longrightarrow x_2' = a_2x_2 + b_2$$
$$x_3 \longrightarrow x_3' = a_3x_3 + b_3 \longrightarrow \text{Is } x_1'+x_2'+x_3'+x_4'+x_5'+ > 0 ? \xrightarrow{\text{Yes}} \boxed{y = 1}$$
$$x_4 \longrightarrow x_4' = a_4x_4 + b_4 \xrightarrow{\text{No}} \boxed{y = 0}$$
$$x_5 \longrightarrow x_5' = a_5x_5 + b_5$$

And we're not finished yet!

The output, **y**, might be a different function of the inputs, not just this zero or one function.

And there might be more than one output, the output might also be a vector, $y_1, y_2, \ldots$

And ... there might be more than one layer of calculations ...

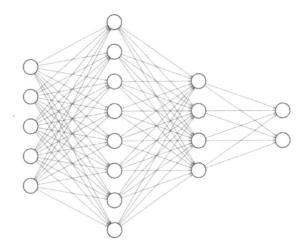

This schematic diagram shows a more typical neural network, but even this one is relatively simple compared to those used in practice. In this example, there are five inputs, on the left, representing a five-dimensional feature vector. As you go along an arrow you multiply by some parameter, and then add another parameter. When you get to a circular cell you add up all the numbers coming in from the arrows on the left and perform some function on it. That function could be the brain-like zero/one or something else. Then the resulting numbers leave the cell along other arrows where the process is repeated. Eventually, having passed through, in this example, two 'hidden' layers, there are two outputs. In the first hidden layer here there are eight nodes, and four in the second.

That's quite exhausting! But I don't expect you to understand it all at first reading. And I'm certainly not expecting you to do this in the Project!

In the above, I mentioned parameters a few times. With the neural network architecture (the number of cells, hidden layers, etc.) in this diagram here there would be getting on for 200 parameters. These would all typically be determined by minimizing a cost function. Once you have more nodes, more hidden layers, more outputs, you can see how easy it is to have neural networks with millions or billions of parameters. For comparison, the human brain has 60 trillion, that's 60,000,000,000,000, connections!

A Real Example

For my real example, I am going to use a neural network to recognize handwritten numbers, from 0 to 9. This example is commonly found in textbooks, because of the ease of access to data and it is relatively straightforward to implement … if you are a programmer. There is sample computer code for this all over the internet. But it's still far too advanced for this book!

So how do I input handwriting into the computer? Images need to be digitized. For example, turn each greyscale image into a square of 28 x 28 pixels. Each pixel is given a number from zero to 255, zero being pure white, 255 being the blackest black. Each handwritten digit is then represented by a vector of dimension 28 x 28 = 784. That's our input vector, much bigger than anything we've seen so far.

For training the algorithm and optimizing to find the tens of thousands of parameters we need data. Fortunately, there is a very famous data set, the

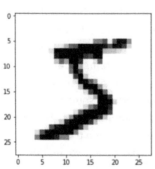

Modified National Institute of Standards and Technology (MNIST) database of 60,000 training images and 10,000 testing images. These are samples handwritten by American Census Bureau employees and American high school students. In the Data section at the end of the book I show where you can find this data.

Here's one example of what one such digitized image looks like. And its raw data.

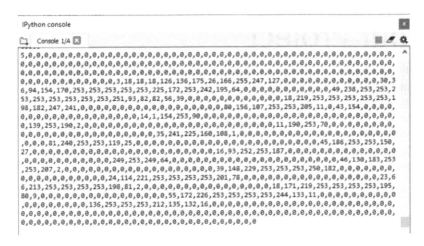

The network that I used had 784 inputs, one hidden layer with 50 nodes, and 10 outputs. Why 10 outputs and not just the one? After all we are trying to predict a number. I've explained this before, as a drawing it is not possible to say that, say, a 7 is mid-way between a 6 and an 8. This is a *classification* problem.

Having trained the algorithm, I then showed it a few of my handwritten digits. I showed it my number '3,' as you've seen on page 27. The neural network got this one right, it knew it was probably a three.

But then I showed it my seven.

It got this one wrong. Perhaps because it's European style, with the bar.

Because the output from the network is a ten-dimensional

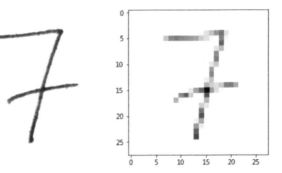

vector, we can analyse what the algorithm thought of my seven. It reckoned that there was an 84.2% probability of it being a two, just 6.8% of being a seven, and 5.7% of it being a number four. But that's my handwriting. On test data the algorithm had a 95% success rate.

Your Project

Building your own neural network is not easy. So instead for your project I'm going to give you three ready-made neural network algorithms and you have to figure out what the results represent. I will give you a few clues.

They all have two-dimensional input vectors and output a single number.

1. The first algorithm is very easy. It doesn't even have any hidden layers. It looks like this:

$x_1 \longrightarrow x_1' = 0.6x_1$

$x_2 \longrightarrow x_2' = 0.6x_2 - 1$

Is $x_1' + x_2' > 0$?

Yes $\rightarrow$ $y = 1$

No $\rightarrow$ $y = 0$

Try inputting just zeroes and ones for the x_1 and x_2, so that's four different combinations. Present the results in a grid.

Clue: This is a very special function in the world of logic.

x_1	x_2	y
0	0	
1	0	
0	1	
1	1	

2. The second algorithm is very similar:

$x_1 \longrightarrow x_1' = 1.5x_1$

$x_2 \longrightarrow x_2' = 1.5x_2 - 1$

Is $x_1' + x_2' > 0$?

Yes $\rightarrow$ $y = 1$

No $\rightarrow$ $y = 0$

Again, input just zeroes and ones and put the results into a similar grid.

Clue: Another special logic function.

x_1	x_2	y
0	0	
1	0	
0	1	
1	1	

3. The third algorithm is a bit harder. This is yet another special logic function but this time one which cannot be represented by a neural network without a hidden layer. That's fine, because in practice almost all neural networks have hidden layers, sometimes very many of them.

Hidden Layer!

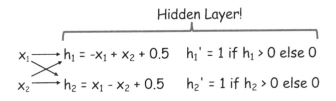

$x_1 \longrightarrow h_1 = -x_1 + x_2 + 0.5 \qquad h_1' = 1$ if $h_1 > 0$ else 0

$x_2 \longrightarrow h_2 = x_1 - x_2 + 0.5 \qquad h_2' = 1$ if $h_2 > 0$ else 0

Is $-h_1' - h_2' + 1.5 > 0$?

Yes $\longrightarrow y = 1$

No $\longrightarrow y = 0$

For this example, you might want to fill out a bigger grid to help you.

x_1	x_2	h_1	h_2	h_1'	h_2'	$h_1' + h_2'$	y
0	0						
1	0						
0	1						
1	1						

What is this special logic function?

And what is different about the third example, what makes it impossible to represent the function without using a hidden layer? To see this, it might help if you draw a graph with axes x_1 and x_2, draw a dot for each of the four combinations and colour the dots as red for $y = 0$ and blue for $y = 1$.

IN DEPTH: REINFORCEMENT LEARNING

I have saved what might be the most interesting methodology to last.

Reinforcement learning is one of the main types of machine learning. Using a system of rewards and punishments, an algorithm learns how to act or behave. It might learn to play a game or move around an environment. The key to reinforcement learning is that the game or environment is not explicitly programmed. This means that the algorithm has to learn the consequences of taking actions from taking those actions. The algorithm learns by trial and error.

While supervised learning is about deciding what something is, and unsupervised learning is about finding relationships in data, reinforcement learning is about teaching the machine to do something. And it really is inspired by behavioural psychology. We want an algorithm to learn what actions to take so as to maximize some reward (and/or minimize punishment). Because you want the machine to learn to attain some goal it is very common to see the method used for playing games. And our examples here will also be from, or related to, games. But the trick with reinforcement learning is that you don't necessarily tell the machine explicitly what the goal of the game is or even what the rules are. It will learn by trial and error as it interacts with its environment.

You can see this happening if you give a two-year old the TV remote control. They've seen you use the remote control to switch on the TV and change channels so they know of some link there. But when first given the remote they will press buttons at random until something happens. There will be a phase during which they home in on the most important buttons, and after a while they learn which is the button to switch the TV on, to change channels, to start a DVD. And thereafter they will use a subset of buttons reliably. Funnily though they never seem to learn how to switch the TV off, not until their mid forties.

Another way of looking at trial and error is in terms of exploiting and exploring. On one hand you want the machine to take advantage of/exploit everything it has learned in order to win the game, say, but on the other hand it won't learn anything unless it has done plenty of exploration beforehand. Without exploring, the machine can get stuck in a rut, taking actions that are less than optimal. Getting a balance between exploiting and exploring will be important to our learning algorithms.

What It Is Used For

- Playing games: Reinforcement learning agents can learn to play complex games such as Go, chess, or Atari by exploring the game environment and maximizing their rewards based on their actions
- Robotics: Reinforcement learning agents can learn to control robots for industrial automation, such as assembling parts, sorting objects, or navigating obstacles
- Business strategy: Reinforcement learning can be used to optimize business decisions, such as pricing, inventory management, marketing campaigns, or customer service
- Education: The algorithms can learn to create personalized learning systems that provide customized instruction and materials according to the needs and preferences of students
- Data-center cooling: Reinforcement learning agents can learn to control the cooling system of a data center by predicting the future energy consumption and minimizing the power usage while maintaining safety standards
- Self-driving cars: Learning agents can learn to drive autonomously by sensing the road conditions, traffic signals, pedestrians, and other vehicles, and taking actions that ensure safety and efficiency
- Finance: Reinforcement learning agents can learn to trade stocks, currencies, or cryptocurrencies by analyzing the market trends, risks, and opportunities, and maximizing their profits

The Details

A good way to explain how reinforcement learning works is via the jargon. I haven't put this jargon into the earlier chapter since it is unique to reinforcement learning. And while I explain the basic jargon I will refer to some common games.

Action: What are the decisions you can make? Which one-armed bandit do you choose in a casino? (We'll do this in detail soon.) Where do you put your cross in the game of Noughts and Crosses? (Mentioned earlier.) Given a choice, which card should you play next in a round of Bungo? Those are all actions.

Reward/Punishment: You take an action and you might get rewarded. You press a button on the wall in Doom and the BFG is revealed. You take your opponent's piece in checkers. It's white chocolate and you eat it. Those are examples of immediate rewards. But there might not be any reward until the end of the game. At the end of the chess game the winner gets the $1,000 prize. But there aren't just rewards. To win the prize you have to be the first to solve the jigsaw puzzle. Every second you take can be thought of as a punishment.

State: The state is a representation of how the game is now. Think of it as a snapshot of the gameboard, for example. It might be the positions of the Os and Xs in a game of Noughts and Crosses. Or the positions of the stones in a game of kōnane. The state is an interesting concept. How much information is needed to represent a state? In Noughts and Crosses, Donald Michie needed 304 matchboxes.

In Blackjack you need at a minimum to know the count of the cards you hold, and how many Aces, and what is the dealer's upcard. To stand a chance of winning in a casino you also need to know something about what cards have been dealt from the deck(s). But sometimes the amount of information you must store to represent the state is prohibitively large. In Go there are typically 361 points, each of which could be empty, or be occupied by a white or black stone. So you might expect there to be 3^{361} possible states. But because not all states are legal the correct number is merely 208,168,199,381,979,984,699,478,633,344,862,770,286,522,453,884,530,548, 425,639,456,820,927,419,612,738,015,378,525,648,451,698,519,643,907,259, 916,015,628,128,546,089,888,314,427,129,715,319,317,557,736,620,397,247, 064,840,935.

A Real Example

You are far too young to be visiting a casino. You may even be too young to go into an amusement arcade at the end of Brighton pier, or, less glam, in a service station on the M4 motorway. So you might not know about the horror of the one-armed bandit or fruit machine. Why I call this a horror I will explain in a moment.

The way this machine works is that you put the money you earned from your 5 a.m. newspaper round into a slot in the machine, then press a big button (or, in ye olde casinos of yore, pull on a lever, hence the 'one arm,' the 'bandit' bit will be clear soon). Wheels spin before your eyes and three or more symbols settle and appear in a row. Traditionally these were images of fruit (hence ...). If these three or more symbols are among certain special combinations, three cherries I recall being particularly favourable, then money pours out of a slot. You have won. Casinos will lure you into this suckers' game by announcing how much money can be won in the jackpot.

Unfortunately, the odds are against the punter. Typically for every dollar you put into the machine you will be lucky if you get back 95 cents, it might even be as low as 70 cents. You will lose your money very quickly. Hence the earlier 'horror' and the name 'bandit.' Just look at the people playing these machines, do you aspire to be like them?

[Editor: Enough with the moralizing, Wilmott.]

Ok, let's see why this is a good exercise for reinforcement learning.

Suppose we are sitting in front of a row of 10 fruit machines in a casino in Las Vegas. It's 7 a.m. on a Wednesday in February. (It should be quiet then.) And I'm going to let you play the bandits for free! You are going to pick one bandit, pull the arm, and wait for the result. You can then try another bandit, or stick to the same one. You will do this over and over trying to win as much as possible.

But there's something about these machines that I'm not going to tell you. So look away now, don't read the next bit. Ok?

Some of these machines have better odds of winning than others! Here's a list of the ten Bandits and the probability of winning for each one.

Clearly Bandits 1 and 5 are pretty rubbish, but Bandit 4 is the best of the lot.

Unfortunately for you, you don't know what these probabilities are, you have no clue as to which is the best bandit. You don't even know that they have different probabilities. That's the idea behind reinforcement learning, that you don't necessarily know the rules of the game, what action leads to the best outcome, you just have to experience playing the game and learn from the results.

Bandit 1	10%
Bandit 2	50%
Bandit 3	60%
Bandit 4	80%
Bandit 5	10%
Bandit 6	25%
Bandit 7	60%
Bandit 8	45%
Bandit 9	75%
Bandit 10	65%

Ok, you can look again.

Let's play!

You pick Bandit 2, pull the lever and ... you lose. Sorry!

Try again.

Not having had any luck with Bandit 2, you now try Bandit 3. You win! Congratulations.

Feeling lucky? Ok, you try Bandit 3 again ... and lose.

What now?

To be fair, you haven't got much information to go on, have you. Three pulls is not exactly statistically significant.

Ok, you try ... you try ... Bandit 10. You win.

Let's continue with Bandit 10. You win again!

Onto a good thing here, try Bandit 10 again. You lose. Darn it.

You keep trying Bandit 10 and you do quite well. But you can't help wondering whether other Bandits are even better. So you experiment, you play different bandits.

As you do this you are *experiencing* the percentages.

You are *learning* about the odds.

And you are being *rewarded*, in cash.

This example shows you the basics of reinforcement learning and how it might be programmed in a computer as an example of machine learning. It is also a particularly simple example because there are no *states* to worry about. Actions, yes, which bandit to pull. But no states.

Part of the algorithm is to assign a 'value' to each bandit, or value of the action of pulling the bandit. In this example, as we play each bandit we just keep track of how many times we have won as a fraction of the number of times we've played that bandit. That's the value in each bandit.

Start by picking a bandit at random. Play. Win or lose. Calculate the value.

After each pull we have to choose the next bandit to try. This is done by most of the time choosing the *action* (the bandit) that has the highest *value* at that point in time. But every now and then, let's say at a random 10% of the time, we simply choose an action (bandit) at random with all bandits equally likely.

On the next page there is an example of the possible sequence of events, just showing the first few choices of bandit. When working through this example remember how there are just two ways to choose which bandit to pull:

1. Pick a bandit that has the highest value at that time, with the value being the ratio of wins to number of pulls *for that bandit*
2. Pick a bandit at random. This will happen 10%, say, of the time even if you end up picking a bandit that looks not so good

It will take a lot of pulls, but eventually this algorithm will home in on Bandit 4, it will eventually have the highest value and keep being picked, almost all of the time.

'Almost all' of the time because of that random exploration. The reason for every now and then picking a bandit at random is so we don't get stuck in a good but sub-optimal rut.

Notes	Bandit picked	Win or Lose	No. of pulls (for that bandit)	No. of wins (for that bandit)	Value (for that bandit)
Pick at random, #2	2	L	1	0	0
Pick at random, #3	3	W	1	1	1
Stick with #3, highest value	3	L	2	1	0.5
Stick with #3, highest value	3	L	3	1	0.333
Pick at random, #10	10	W	1	1	1
Stick with #10, highest value	10	W	2	2	1
Stick with #10, highest value	10	L	3	2	0.667
Stick with #10, highest value	10	L	4	2	0.5
Stick with #10, highest value	10	L	5	2	0.4
Pick at random, #5	5	L	1	0	0
Back to #10, highest value	10	W	6	3	0.5
Stick with #10, highest value	10	W	7	4	0.571
Stick with #10, highest value	10	W	8	5	0.625
...	...	...	...	...	...
Pick at random, #4	4	W	1	1	1
Stick with #4, highest value	4	W	2	2	1

Your Project

I want you to find your way out of a maze! It's not the most challenging of mazes, look at the picture. You start in Cell A1, move from cell to cell. When you reach C3 you have finished. How many steps will it take?

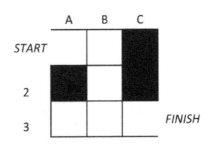

(It's not as easy as it looks. Imagine that you are in a basement, no lighting, there are doors you can go through, but no idea how many rooms there are. That's what reinforcement learning is all about. The rules of the game aren't explicitly programmed, you just have to play the game and see how you get on!)

You are going to start out by moving around the maze at random.

Get a six-sided die. You will roll this to determine what path to take, at least to start with.

1. **A1**: There's only one way to go from Cell **A1**, so no roll is needed
2. **B1**: There are two ways to go, back to **A1** (roll a 1, 2 or 3) or down to **B2** (roll a 4, 5 or 6)
3. **B2**: Roll a 1, 2 or 3 to go back up to **B1**, 4, 5 or 6 to go down to **B3**
4. **B3**: There are three cells you can go to from here. Roll 1 or 2 and you go to **B2**, 3 or 4 and it's **A3**, 5 or 6 and it's **C3** and you are out of the maze
5. **A3**: No roll needed, you can only go to Cell **B3** from here

I think you'll find that on average it will take you a surprising number of moves to get from **A1** to **C3**. If my calculations are correct, you can expect to take 18 moves, thanks to a lot of toing and froing. Even to get from **B3**, which is just one move away from the exit, will take an average of nine random rolls! This is far from optimal.

To optimize we need to keep track of another value function. But unlike the earlier one-armed bandit example, here we need to value not just the action, but also the *state*. By 'state' I simply mean which cell are we in currently, and by 'action' I mean which way do I move from that state.

I have sketched out the sort of experimental data you can write down to help with this.

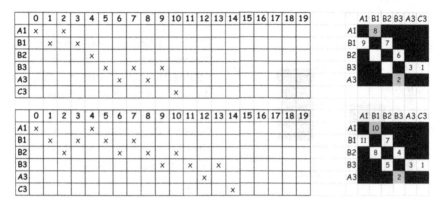

Interpret this figure as follows. Here are two goes at escaping the maze. I want you to do a dozen like this. (And with harder mazes, a lot more.) The grids on the left show you the escape routes taken thanks to the random rolling of the die. The first, top, route was

$$A1 \rightarrow B1 \rightarrow A1 \rightarrow B1 \rightarrow B2 \rightarrow B3 \rightarrow A3 \rightarrow B3 \rightarrow A3 \rightarrow B3 \rightarrow C3$$

That's 10 moves.

The second attempt took 14 moves.

The grids on the right-hand side show how long it took to get out of the maze given which cell you were in and which cell you went to. This is the value function I mentioned. The black cells in those grids are just actions that are not possible. You cannot go from B2 to A1 for example. The grey cells are ones for which there is no choice. From B1 and B2 there are two options, from B3 there are three.

We put numbers into the grid as follows. See how there are two escapes starting at A1 in the top route? That's because the route backtracked, once starting at A1 and then going through A1 a bit later. So it took 10 or eight moves from A1 to C3. Which was faster? Eight, obviously. So put eight in the grey cell. Similarly there were two occasions when the route took us from B3 to A3, with five or three moves to escape. So put three in the B3/A3 cell in the grid. And so on. There was no path in which we went from B2 to B1 so don't put anything in that cell. (Although if you programmed this up you might put in a default really large number, otherwise the algorithm might interpret an empty cell as taking zero steps to escape!)

After you've done a few of these random routes you'll be able to see which moves are optimal, giving the quickest escape from the maze.

The final part of this project is to work in pairs. One person designing a maze, the other trying to get through it. Obviously the second person doesn't get to see the maze. All they see would be instructions along the lines of:

> From Apple you can move to Banana or Clementine
>
> From Banana you can move to Apple, Clementine, Date or Elderberry.
>
> From Clementine you can move to etc.

I've replaced all the obvious $A1$s, etc. with fruit names so as to make it harder for anyone to visualize the layout of the maze. Start with a five-by-five maze to make things just the right amount of challenging. The maze solver can try moving through randomly, rolling a die perhaps, or just guessing. The larger the maze the more attempts they'll have to make before they start to home in on the optimal route.

With more complicated mazes and other reinforcement learning problems you will probably have to be a bit subtle in determining what is optimal. You would still build up a value function representing the rewards for taking certain actions in each state but you would also throw in the occasional random actions, just in case you've missed something better. It is no different from

finding the best route home. You might think you have the best route but every now and then try a different path. When doing the project in pairs, the maze setter could have one obscure route that would not be found if one is too keen to follow the well-trodden paths.

Here's a maze with a very quick route to the exit. But it's via wormhole that is only open every sixth move!

EPILOGUE

It was great fun writing this book, and I hope fun for you reading it. And maybe even educational! I did cheat and use some AI, in the 'What It Is Used For' parts. But apart from those the rest of the words are my own. I take full responsibility for them!

Please let me know of any improvements that I can make to this book. And let me know how you got on with the Projects. (Also, if you want the Official Wilmott Rules of Bungo!) I can be contacted at paul@wilmott.com. I'm surprisingly approachable.

(I don't have Twitter, FaceBook, Instagram, etc. accounts ... you never know how much data They are collecting on you!)

DATA

For proper training of many algorithms, you need plenty of data. While you can always take surveys of people's heights, weights, pet ownership, etc. you will start to struggle when it comes to property values, traffic statistics, and so on. You'll need to get data from the internet. Good places to start looking for data are https://toolbox.google.com/datasetsearch and https://www.kaggle.com. The latter run competitions in AI and you'll need to register with them.

Here are some datasets you might want to download to play around with, to get you started.

Titanic: A very popular dataset for machine-learning problems is that for passengers on the Titanic, and in particular survivorship based on information such as gender, class of ticket, etc. Go to https://www.kaggle.com/c/titanic.

Irises: There is a famous dataset for measurements of varieties of irises. That's the flower, not the coloured bit in your eye. This dataset can be found everywhere, for example https://www.kaggle.com/uciml/iris.

Heights and weights: Ten thousand heights and weights of adults can be found at https://www.kaggle.com/mustafaali96/weight-height.

Handwriting: Seventy-thousand handwritten images of digits can be found at http://yann.lecun.com/exdb/mnist/. These are representations of the images written by employees of the American Census Bureau and American high school students in CSV format.

House of Commons Voting: The site https://www.publicwhip.org.uk collects data for activity in the UK's Houses of Commons and Lords. Dig down to https://www.publicwhip.org.uk/project/data.php and https://www.publicwhip.org.uk/data/ to get at the raw data.

Mushrooms: Fancy predicting whether a mushroom is edible or poisonous? Well, pop over to https://www.kaggle.com/uciml/mushroom-classification. This contains data for over 8,000 mushrooms with information about their geometry, colours, etc. and whether they are edible. Rather you than me. (Remember what it says right at the start of this book? "The publisher and the author of this book shall not be liable in the event of incidental or consequential damages in connection with, or arising out of reference to or reliance on any information in this book or use of the suggestions contained in any part of this book.")

Financial: Plenty of historical stock, index and exchange rate data can be found at https://finance.yahoo.com. For specific stocks go to https://finance.yahoo.com/quote/[PUTTHESTOCKSYMBOLHERE]/history/. The Bank of England has economic indicators, interest rates etc. at https://www.bankofengland.co.uk/statistics/.

Banknotes: Data for characteristics of banknotes can be found here http://archive.ics.uci.edu/ml/datasets/banknote+authentication.

Animals: Data for characteristics of various animals can be found here http://archive.ics.uci.edu/ml/datasets/zoo.

Actors: Popular actors can be found here https://today.yougov.com/ratings/entertainment/popularity/all-time-actors-actresses/all.

INDEX

Page numbers in **bold** are major entries about the subject.

ACKNOWLEDGMENTS

I would like to thank profusely my favourite *middle* schooler, Genevieve Wilmott, and mathematics teacher Heerpal Sahota, at Latymer Upper School, for their suggestions.

I am grateful to the Estate of Donald Michie for giving permission for me to use two photographs.

Thank you to Bing for its photo-generation skills. The cover was designed by the talented Liam Larkin using images from Freepik.com.

Finally, I asked ChatGPT to write an acknowledgment to itself. This is what it said:

> Something went wrong. If this issue persists please contact us through our help center at help.openai.com.

I think that's a good place to call it a day.

ABOUT PAUL WILMOTT

Paul is a practising mathematician. He researches in mathematics, teaches mathematics, and writes about mathematics.

His love of mathematics started at Stanton Road County Primary School in Bebington, Merseyside. He fondly remembers his first ever homework, pages and pages of manipulation of fractions. Wirral Grammar School for Boys followed, where he took as many mathematical subjects as possible. He then studied mathematics at St Catherine's College, Oxford, where he also received his doctorate.

Paul was briefly a professional juggler with the Dab Hands troupe, and was an undercover investigator for Channel 4. He also has three half blues from Oxford University for Ballroom Dancing. And he plays the 'ukulele.

Made in the USA
Las Vegas, NV
11 June 2024

90977094R00057